Canada: The State of the Federation 2012

Regions, Resources, and Resiliency

Edited by
Loleen Berdahl,
André Juneau, and
Carolyn Hughes Tuohy

Institute of Intergovernmental Relations
Queen's Policy Studies Series
School of Policy Studies, Queen's University
McGill-Queen's University Press
Montreal & Kingston • London • Ithaca

The Institute of Intergovernmental Relations

The Institute is the only academic organization in Canada whose mandate is solely to promote research and communication on the challenges facing the federal system.

Current research interests include fiscal federalism, health policy, the reform of federal political institutions and the machinery of federal-provincial relations, Canadian federalism and the global economy, and comparative federalism.

The Institute pursues these objectives through research conducted by its own associates and other scholars, through its publication program, and through seminars and conferences.

The Institute links academics and practitioners of federalism in federal, provincial, and territorial governments and the private sector.

The Institute of Intergovernmental Relations receives ongoing financial support from the J. A. Corry Memorial Endowment Fund, the Royal Bank of Canada Endowment Fund, and the governments of Manitoba and Ontario. We are grateful for this support, which enables the Institute to sustain its program of research, publication, and related activities.

L'Institut des relations intergouvernementales

L'Institut est le seul organisme universitaire canadien à se consacrer exclusivement à la recherche et aux échanges sur les enjeux du fédéralisme.

Les priorités de recherche de l'Institut portent présentement sur le fédéralisme fiscal, la santé, la modification des institutions politiques fédérales, les mécanismes des relations fédérales-provinciales, le fédéralisme canadien dans l'économie mondiale et le fédéralisme comparatif.

L'Institut réalise ses objectifs par le biais de recherches effectuées par des chercheurs de l'Université Queen's et d'ailleurs, de même que par des congrès et des colloques.

L'Institut sert de lien entre les universitaires, les fonctionnaires fédéraux, provinciaux et territoriaux et le secteur privé.

L'Institut des relations intergouvernementales reçoit l'appui financier du J. A. Corry Memorial Endowment Fund, de la Fondation de la Banque Royale du Canada, et des gouvernements du Manitoba et de l'Ontario. Nous les remercions de cet appui qui permet à l'Institut de poursuivre son programme de recherche et de publication ainsi que ses activités connexes.

ISSN 0827-0708
ISBN 978-1-55339-210-1 (pbk.)
ISBN 978-1-55339-211-8 (epub)
ISBN 978-1-55339-212-5 (pdf)

CONTENTS

PREFACE

The Institute of Intergovernmental Relations has a tradition of entering into partnerships for many of its projects. In this instance, we worked with a group of senior scholars at the University of Toronto. The theme for this book emerged in early 2012 from exchanges and conversations triggered by Richard Simeon following the opinion of the Supreme Court of Canada on a national securities regulator. As Richard, David Cameron, Carolyn Hughes Tuohy, and I contemplated holding a conference on the implications of the Court's opinion, we realized that we needed to look beyond that relatively narrow issue to consider what Richard called the changing landscape of Canada's political economy. Carolyn Tuohy then agreed to "represent" her colleagues in designing the conference and the publication, and the University of Toronto School of Public Policy and Governance became a co-sponsor of the event. It is fitting that the book that results from those discussions and the conference is dedicated to Richard Simeon, the outstanding scholar of federalism and former director of the Institute, who passed away in October 2013. The dedication follows this preface.

I am pleased that we avoided an obvious central Canada bias by asking Professor Loleen Berdahl, of the University of Saskatchewan, to be the lead on the conference and the lead editor of this book with the support of Carolyn Tuohy and myself. I thank both of them for being such good partners.

Regional resource disparities and the tensions they generate are a perennial Canadian topic. Governments, unfortunately but not surprisingly, are reluctant to tackle these issues. This imposes an even greater responsibility on universities and think tanks to study and understand these issues and to disseminate the work of scholars whose chapters are in this book. On behalf of the editors, I would like to thank the authors for their contributions and the anonymous reviewers who commented on selected chapters. We also want to thank other participants who contributed much to our conference as chairs or as panelists: André Pratte, Greg Marchildon, Andrew Coyne, André Plourde, Roger Gibbins, Elizabeth Beale, David Cameron, George Anderson, Jim Carr, Don Drummond, Keith Banting, and Dylan Jones.

I want to thank our sponsor, COGECO, for a second year in a row.

Finally, we want to thank Mary Kennedy for, as usual, her invaluable assistance in conference organization and all matters of follow-up, and Ellie Barton, Valerie Jarus, and Mark Howes, from the publications unit of the School of Policy Studies, for their work on this volume.

André Juneau
Director

RICHARD SIMEON (1943–2013)

Most readers of a book on the state of the federation will know, sadly, that Richard Simeon passed away on October 11, 2013. Richard was one of Canada's most distinguished political scientists, and his fellow students of federalism are proud that his well-deserved reputation came about largely because of his work on federalism in Canada and abroad.

The editors of this book, Loleen Berdahl, Carolyn Hughes Tuohy, and I, have dedicated this book to Richard.

Richard Simeon was recruited to Queen's University in 1968 by Professor John Meisel to teach political science. He was already famous for his Yale University PhD dissertation on federal-provincial diplomacy, a phrase he coined. He became the director of the Institute of Intergovernmental Relations in 1976. He wrote recently that Ron Watts had given him the best job of his life.[1] As director for seven years, he guided the Institute to play an influential role during one of the most difficult periods for Canadian national unity. Working with his existing network and creating new ones, he was tireless in seeking to promote an understanding both of the tensions and of the potential avenues for the future of the country.

From 1985 to 1991, Richard was the director of the School of Public Administration (now known as the School of Policy Studies). This appointment reflected another one of his deep interests, the making of public policy. A central question for him was, "How does federalism matter for equality, for social justice, for addressing contemporary challenges in a timely and effective fashion?"[2] This of course remains more than ever a crucial question.

In 1991, Richard left Queen's and went to the University of Toronto, but he continued to be a friend and an advisor to the Institute. His role in the design of the 2012 State of the Federation conference is outlined in the preface to this book.

Throughout his academic career, Richard Simeon remained a public intellectual in the best meaning of that term. He made important contributions to public policy

---

[1] "Reflections on a Federalist Life," in *The Global Promise of Federalism*, eds. Grace Skogstad, David Cameron, Martin Papillon, and Keith Banting (Toronto: University of Toronto Press, 2013), 279.

[2] Ibid., 284.

and federalism by advising governments in Canada and abroad in a wide range of ways. To name just a few, he was a research coordinator for the Macdonald Commission on the Economic Union and Development Prospects for Canada, he was one of Premier Peterson's advisors on the Meech Lake Accord, he worked with Al Johnson in South Africa and, through the Forum of Federations, in many other countries in the developing world.

We will miss him and take comfort in the ongoing contributions to the study and practice of federalism by his many students.

André Juneau

CONTRIBUTORS

Daniel Béland holds the Canada Research Chair in Public Policy (Tier 1) at the Johnson-Shoyama Graduate School of Public Policy (University of Saskatchewan campus).

Loleen Berdahl is an associate professor of political studies at the University of Saskatchewan.

Douglas Brown is associate professor in the Department of Political Science at St. Francis Xavier University. He is a fellow and former executive director (1993–96) of the Institute of Intergovernmental Relations.

Serge Coulombe is professor of economics at the University of Ottawa, his base since 1982. His recent research on the Dutch disease has been at the centre stage of the Canadian economic policy debate. In 2009, he won the University of Ottawa President's Award for Media Relations.

Thomas J. Courchene is an adjunct professor in the School of Policy Studies, Queen's University and a former director of the Institute of Intergovernmental Relations in the Queen's School of Policy Studies. He is also senior scholar at the Institute for Research on Public Policy (Montreal).

Monica Gattinger is associate professor at the University of Ottawa's School of Political Studies and chair of the university's Collaboratory on Energy Research and Policy.

Frank L. Graves is the founder and president of EKOS Research Associates Inc. He is one of Canada's leading experts on public attitudes and public policy. He was recently awarded a fellowship in the Market Research and Intelligence Association, its highest award for lifetime achievement.

André Juneau is a fellow of the Institute of Intergovernmental Relations at Queen's University. He served as the Institute's director from March 2010 to December 2013. In a 34-year career in the Government of Canada, his main interests were social policy and intergovernmental relations.

André Lecours is professor in the School of Political Studies at the University of Ottawa.

Christian Leuprecht is associate dean of the Faculty of Arts and associate professor of political science at the Royal Military College of Canada. He is cross-appointed to the Department of Political Studies and the School of Policy Studies at Queen's University, where he is also a fellow at the Institute of Intergovernmental Relations.

Geneviève Motard is professor of constitutional and aboriginal law at the Faculty of Law, Université Laval, Québec.

Tim Nieguth is associate professor in the Department of Political Science, Laurentian University.

Tracey Raney is associate professor in the Department of Politics and Public Administration at Ryerson University. In 2013, she was awarded the Jill Vickers prize for the best paper presented on gender and politics at the 2012 Canadian Political Science Association annual meeting.

Jeff Smith is a senior consultant with EKOS Research Associates Inc. A seasoned public opinion researcher, he has devoted much of his career to asking Canadians about their views on everything from the economy to health care to security.

Carolyn Hughes Tuohy is a professor emeritus of political science and senior fellow in the School of Public Policy and Governance at the University of Toronto. She is a fellow of the Royal Society of Canada.

Michael Valpy is a journalist, senior fellow at Massey College, and fellow of the School of Public Policy and Governance at the University of Toronto.

Donna E. Wood is an adjunct assistant professor in the Department of Political Science at the University of Victoria. She is co-editor of the June 2013 special issue of *Canadian Public Administration,* "Comparing Modes of Governance in Canada and the European Union: Social Policy in Multilevel Systems."

I

Introduction

REGIONS, RESOURCES, AND RESILIENCY: INTRODUCTION AND OVERVIEW

Loleen Berdahl, Carolyn Hughes Tuohy, and André Juneau

As recent Canadian debates about resource development, "Dutch disease," and employment insurance demonstrate, regional tensions remain alive and well in the Canadian federation. Past regional disputes often centred on questions of federal government "fairness" to particular provinces, most notably Quebec. The outcome of the 2012 Quebec election has not yet significantly reopened and broadened those debates, presumably because of the challenges faced by the minority government. In any event, the economic, political, and social implications of the unequal distribution of natural and human resources among provinces will remain. In the current market, natural resource–rich provinces, including Alberta, Saskatchewan, and Newfoundland and Labrador, enjoy fiscal capacities far superior to other provinces, and wrestle with labour market supply challenges. At the same time, provinces with fewer natural resource industries, including Ontario and Quebec, struggle with manufacturing industry decline and relatively high unemployment. Such differences lead to (often heated) discussions about the impact of natural resource economies on the value of the dollar, the appropriate model for employment insurance, and the federal equalization program, among other issues.

The State of the Federation 2012 conference, held in Kingston on November 30 – December 1, 2012, brought together academics, policy-makers, and politicians to engage in a constructive dialogue about regionalism, resources, and the resiliency of the Canadian federal system. Questions considered included the following:

- How do provincial and regional differences in economic capacity impact on Canadian federalism? Do the current economic strains represent a unique challenge to Canadian national unity, or do they simply reflect the country's long history of regionalism?

- To what extent do Canada's natural resource industries benefit the Canadian economy? To what extent do they create pressures for other industries?
- Do Canada's federal institutions hinder or promote the ability of the economy to respond to global economic shifts?
- Should Canada pursue national policy approaches, such as a Canadian energy strategy, in areas of provincial or concurrent jurisdiction? If so, what approaches are needed and how can they be achieved?
- Do current intergovernmental structures allow for constructive dialogue about national policy issues? Are other institutional arrangements required?
- Does Canada need new concepts of provincial and regional "fairness" and "equity"?
- What lessons, if any, might be learned from other federal systems? What lessons might be learned from Canada's past?

This volume includes the papers that were presented at the conference. The conference discussants and participants provided valuable feedback to the authors.

While the theme of the conference sought to broaden the debate beyond the consequences of natural resource disparities, in one way or another more than half the chapters touch on energy issues. The Leuprecht, Coulombe, Gattinger, and Courchene chapters are very focused on energy. The chapters by Béland and Lecours, and by Brown, necessarily deal with energy but not as their sole issue. Andrew Coyne's luncheon speech, not reproduced in this book, vigorously challenged the need for an energy strategy.

Functional institutions in a federal or multilevel context are an essential requirement for the effective resolution of issues, a point illustrated by the Wood and Motard chapters. Unexpectedly, the latter chapter, on the Quebec-Cree regional government agreements, previews the theme of the 2013 State of the Federation conference, namely, multilevel Aboriginal governance.

Finally, the chapters by Nieguth and Raney, and by Graves, Smith, and Valpy, remind us that Canada is not only about physical disparities but about common and not-so-common identities and values.

Another way to view this volume is that it continues to assess a question raised in the State of the Federation 2011 conference, namely, whether and to what extent there is and will be, again, a shift in the balance between federal and provincial governments, a shift in influence, and a shift in leadership. Implicit in a number of chapters, and more explicit in the discussion at the conference itself, however, are a set of underlying questions about the relationship of state and society. Throughout our discussions, participants pointed to the importance of trust and moral legitimacy for the operation of our federal institutions. This emphasis on legitimacy raises the question of the extent to which federal institutions are reinforced or, conversely, placed under stress by societal structures. We need, that is, to consider the degree of federalism in Canadian *society*.

- Does the structure of networks, identities, social institutions, and markets make for a "federal society"?
- What social institutions or networks buttress or frustrate structures of federal governance? Have recent developments strengthened or weakened these networks? One of the panelists, Jim Carr, for example, spoke about the process leading to the "Winnipeg consensus" around a national energy strategy, a process that involved representatives from think tanks and business associations. Wood, conversely, noted the withdrawal of federal funding from and consequent weakening of a number of intermediary organizations in the labour market and social services policy arenas.
- What *tensions* exist between market flows and federal institutions and policies, and how can these tensions be managed? Leuprecht's chapter treats the tensions between transborder energy flows and territorial institutions of governance as presenting problems of maintaining institutional equilibria. These problems in turn raise the question of how, if at all, institutions or intergovernmental relations should be adapted to market flows in the energy arena – for example, should they be managed through the development of cap-and-trade regimes involving particular subsets of jurisdictions within and across the Canadian and American federations?
- To what extent will *market and society* responses be able to address the concerns that motivated this conference *without* change in or action through institutions of intergovernmental relations? For example, Elizabeth Beale, in discussing the economies of the Atlantic provinces from the perspective of the Atlantic Provinces Economic Council, suggested that some of the economic stresses within the federation will be eased as firms adapt to changes in terms of trade. Coulombe, however, in his chapter on "Dutch disease," is more circumspect, drawing attention to the costs of such adjustment. At the very least, differences in the terms of trade across provinces will strain the federation in the course of an acute episode of costly adjustment. At worse these differences will result in a chronic problem of Dutch disease.

One real test of the dynamics of state-society relations within the Canadian federation will be the capacity to act on an imperative recognized by all participants at the conference: the need to adopt a price on carbon and to manage disparities in the resulting revenue flows.

II

Regions and Resources: Setting the Stage

2

GO WITH THE FLOW:
THE (IM)PLAUSIBILITY OF A GRAND
CANADIAN INTERGOVERNMENTAL
BARGAIN ON ENERGY POLICY
AND STRATEGY

Christian Leuprecht

The premise of this chapter is the disconnect between energy flows and the systems of governance to which they are subject. Globalization harnesses differentials in the way cities, regions, countries, and continents are endowed with resources. The result is a vast and growing set of real and virtual flows across jurisdictions. Because these flows cross jurisdictional boundaries, their regulation may cause horizontal and vertical collective-action problems among multiple levels of government. The actual and required constitutional powers necessary for effective regulation may be misaligned across jurisdictions and selectively deployed in ways that further exacerbate regional differences. Agnew (1994) refers to the political consequences of this disconnect as the "territorial trap." In Canada, this is particular striking with respect to the production, distribution, and consumption of energy within and across provinces.

Does Canada need a national energy strategy? If so, how much government intervention is desirable to realize that end? And is an intergovernmental bargain even possible, let alone sustainable? Initially the chapter broaches these questions by initiating the reader into the effects that jurisdictional boundaries have on policy. The following section details the particular challenges that federal collective-action problems in Canada raise around energy policy. The next section walks the reader through observations about the intergovernmental dynamics that inform energy policy in Canada. The final section discusses the prospects for forging a national strategy. To ascertain the extent to which Canada may be able to capitalize on the experience of other federations in this policy field, the conclusion situates these implications in a comparative international context.

JURISDICTIONAL BOUNDARY EFFECTS

Space is a way of making sense of the world. Geographical assumptions naturalize the political segmentation of space. The study of intergovernmental relations is particularly afflicted by such assumptions. On the one hand, the hegemonic preponderance of historical institutionalism across the field of intergovernmental relations necessarily causes scholars to gravitate toward the study of institutions, to the detriment of more sociological and ecological perspectives that transcend the institutional explanations. On the other hand, the field of intergovernmental relations is replete with methodological nationalism. By default, its units of analysis are sovereign federal, decentralized and sometimes devolved states, and their semi-autonomous constituent units.

Borders have traditionally been understood "as constituting the physical and highly visible lines of separation between political, social and economic space" (Newman 2006, 144). But their actual significance is found in the bordering process that produces them and the institutions that manage them. These institutions "enable legitimation, signification and domination, [and] create a system or order through which control can be exercised" (Newman 2006, 149). They politicize space and bring it under control. Since the people are, ultimately, sovereign, federalism is sustained by the various governments' accountability to the voters. In a diverse society, however, forging a consensus among voters' expectations is difficult. Canada's inability even to attempt to forge an intergovernmental consensus on energy policy and strategy is, as this chapter will show, a case in point. Indeed, although some might claim that Canada's provincial and national boundaries are little more than arbitrary constructs, these boundaries, their corresponding political institutions, and their territorial priorities, interests, values, and identities weigh heavily on the prospects of achieving a coherent, national intergovernmental energy policy: by mere virtue of different endowment factors, some regions are mainly producers while others are mainly consumers. Quoting Painter (1995, 47): "The state is not only a set of institutions, but a set of understandings – stories and narratives which the state tells about itself and which make it make sense." The emergence of the state has thus been contingent upon certain processes that have turned space into "state space" (Brenner et al. 2003).

Border coefficients to which policy differentials across these sovereign jurisdictions give rise are considerable, and their welfare implications are among the major puzzles in international economics (Obstfeld and Rogoff 2001). Loesch

(1954) in *The Economics of Location* reasoned that, according to neoclassic economics, the borders created by these processes are costly because they are barriers
to free trade and the free flow of goods, labour, or skills. After controlling for
distance and other factors, Engel and Rogers (1996, Table 3, 1117) conclude that
the economic impact of the border on price dispersion across US and Canadian
cities is equivalent to shipping a good 75,000 miles (although Gorodnichenko
and Tesar [2009] subsequently demonstrate that this border effect is entirely
driven by the difference in the distribution or prices within the US and Canada).
McCallum (1995) calculates that the gravity-adjusted volume of trade among
Canadian provinces exceeds provinces' trade with US states by more than a factor of 20. Provincial borders in Canada (Helliwell and Verdier 2001) and state
borders in the United States (Millimet and Osang 2007; Wolf 2000) have a large
and economically significant subnational border effect on decreasing substate
trade flows. Ceglowski (2003) finds that provincial borders in Canada have a
significant impact on intercity price heterogeneity, although the provincial border
effect turns out to be an order of magnitude smaller than the estimates for the
Canada-US border. Contemporary Canadian economist John Helliwell (1998,
2002) has argued that, economic integration notwithstanding, borders continue
to "matter" because they delineate the boundaries of governments. They also
circumscribe social networks and human interactions (Hale and Gattinger 2010).
And, in federations, they reify and institutionalize autonomy with respect to
manifest priorities, interests, and values among jurisdictions.

ENERGY POLICY AS AN INTERGOVERNMENTAL COLLECTIVE-ACTION PROBLEM

Flows affect multiple levels of government and multiple jurisdictions. That raises
collective-action problems in achieving stable, sustainable agreements among parties. Different priorities, interests, and values make it difficult to reach agreement.
In few Canadian policy areas is that more evident than in energy policy. The bulk
of energy infrastructure is either in private hands or owned by Crown corporations
that operate like quasi-private entities. Section 92A of the Constitution Act (1867)
assigns to provincial governments exclusive jurisdiction over non-renewable resources and electricity. Much of the energy infrastructure is subject to provincial
jurisdiction, some to federal jurisdiction, still some, effectively, to both. Although
government may have greater leverage over Crown corporations, in the end, both
Crown corporations and regular private-sector enterprises dealing with energy
have massive capital investments over which government has relatively little
leverage, other than to regulate or provide incentives to spur or discourage certain
kinds of behaviour. The extent of private-sector ownership of critical infrastructure

exacerbates challenges for government to regulate the flows through that infra-structure. In whose interests is government to regulate: consumers or producers? consuming or producing regions? How competing interests are reconciled is at least partially a function of the government's locus of power and political support. In a country where energy policy-making and regulation is relatively decentralized constitutionally, these dynamics are bound to give rise to an array of contentious cleavages that are difficult to reconcile.

Canada has no coherent national energy policy, nor has it ever had one. From the perspective of intergovernmental relations, Canada has never had a grand bargain in this policy area, let alone sustained one. Pierre Elliott Trudeau's National Energy Policy is the *exemple par excellence* of the federal government's attempt to impose a solution top-down that prompted policy failure and regional alienation, precisely because it did not reconcile the interests of producing and consuming regions. Among provinces, however, there is some agreement, both bi- and multilateral. Although energy sources are changing, it used to be that Atlantic Canada heated primarily with oil, Quebec with electricity, and the rest of the country with gas. Ontario relies on nuclear power to generate the majority of its electricity. Ontario's demand for electricity peaks in the summer whereas Quebec's peaks in the winter. Gas pipelines flow from west to east, but also to Canada's west coast as well as southward; incipient efforts are attempting to switch the flow of some pipelines from gas to oil, to reverse the flow of others, and to build or surge capacity going west, south, and east. The extent to which these efforts have thus far been thwarted, notably by Aboriginal and environmental opposition, is a manifest example of intergovernmental collective-action problems (Alternatives North 2008; Angell and Parkins 2011; Bowles and Veltmeyer 2014; Caulfield 2000; Preston 2013; Van Hinte, Gunton, and Day 2007).

The result is a highly variegated system that largely transcends national bound-aries and defies a coherent national strategy. Electricity generation and distribution offers a good example because it is not only an energy source but also a means of transmitting energy, which has the advantage that it can be generated using any number of renewable and non-renewable resources. Figure 1 illustrates why in 2003 a tree branch falling on a transmission line in Ohio caused the lights to go out across the northeast.

Figure 2 illustrates the diversity and degree of variation in fuel options for the generation of electricity across Canada's provinces.

Figure 3 makes explicit interprovincial differentiation and variation in the inte-gration of North America's electricity grid.

Figure 4 depicts the major current and planned oil trunkline network spanning North America.

Figure 1: Integrated North American Transmission Grid

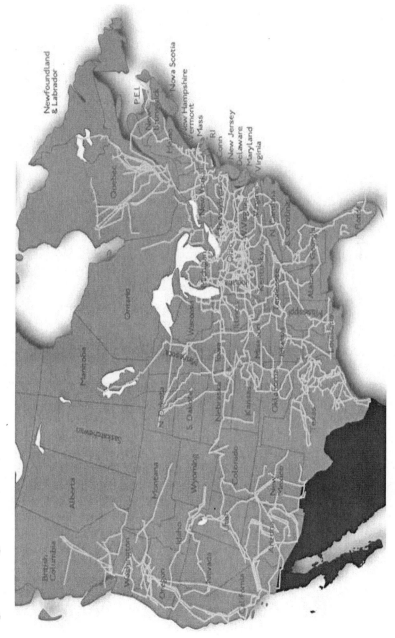

Note: Lines shown are ≥345kV interconnections between Canada and the United States; <345kV lines are not depicted.
Source: Reproduced with permission. Map copyright Canadian Electricity Association (2015, 26).

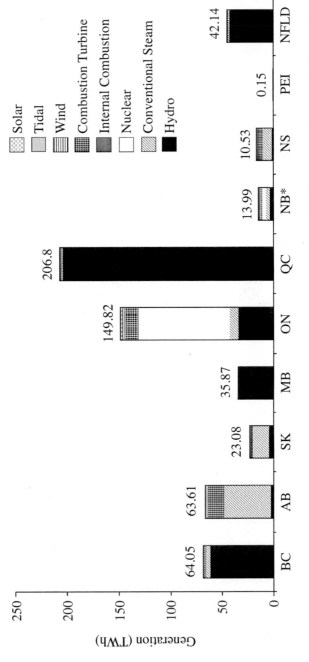

Figure 2: Electricity Generation in Canada by Province and Type of Fuel

Note: *Point Lepreau nuclear generating station resumed power production on November 23, 2012; nuclear is expected to be a major source (about 30 percent) of electricity in New Brunswick.

Source: Reproduced with permission from the Canadian Electricity Association (2015, 16), using Statistics Canada data (2013).

Figure 3: North American Electricity Grid Interconnections

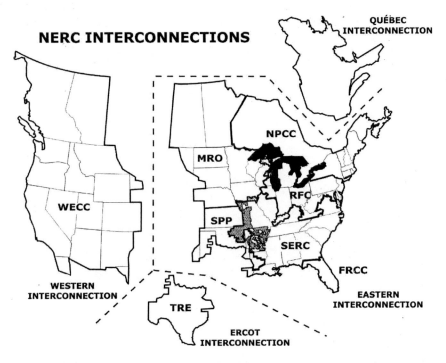

Note: ERCOT = Electric Reliability Council of Texas; FRCC = Florida Reliability Coordinating Council; MRO = Midwest Reliability Council; NERC = North American Electric Reliability Corporation; NPCC = Northeast Power Coordinating Council; RFC = Reliability First Corporation; SERC = Southeast Electric Reliability Corporation; SPP = Southwest Power Pool; TRE = Texas Reliability Entity; WECC = Western Electricity Coordinating Council.

Source: Reproduced with permission from North American Electricity Reliability Corporation. This information from the North American Electric Reliability Corporation's website is the property of the North American Electric Reliability Corporation and is available at http://www.nerc.com/AboutNERC/Documents/Understanding%20the%20 Grid%20DEC12.pdf. This content may not be reproduced in whole or any part without the prior express written permission of the North American Electric Reliability Corporation.

Figure 4: Canadian and US Oil Pipelines

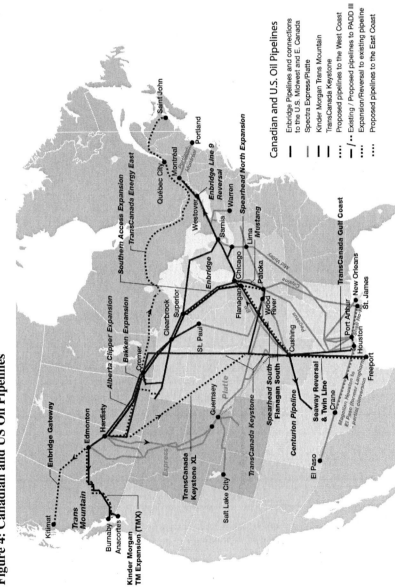

Source: Reproduced with permission from Canadian Association of Petroleum Producers (2014, 22).

INTERGOVERNMENTAL OBSERVATIONS WITH RESPECT TO ENERGY POLICY

First, we observe a regionally segmented policy field in which the major flows run north-south or south-north, thereby traversing national boundaries. Continental integration, however, is characterized by other economic flows and transportation, thus reducing its efficacy as a distinguishing characteristic of energy. Other features of energy are more satisfactory in this regard.

For example, unlike many other economic flows, energy flows are highly oligopolistic, being dominated and directed by the interests of a few very large private-sector players. That is precisely why governments choose to regulate this policy field, to forestall market failure and distortion. Moreover, the regulatory regime for energy flows is best described not as a monopoly, but as an asymmetric duopoly: provinces regulate within their jurisdiction, the federal government regulates interprovincially. Third, energy flows constitute a policy field where development is integral to the economic and fiscal health of both the federal and the provincial governments; it thus tends to trump other regulatory concerns, such as the environment. Furthermore, this is a heavily politicized policy field, one where end users are highly averse to rising prices, most particularly for electricity. Such public resistance, of course, is hardly surprising in a northern democracy characterized by long, cold winters that make heat a basic necessity for survival. When the resulting inelastic demand of end users is coupled with considerable short-term price variability and a very high capital intensity, government intervention to minimize political and economic risks is hardly surprising. Indeed, some subfields depend for their viability on government subsidies that tend to take two forms: having government invest in or underwrite infrastructure, as is the case with nuclear and hydro power; or having government provide direct or indirect subsidies, the most obvious being the lack of insistence on stringent environmental regulation in the case of the tar sands.

Finally, the nature of government investment has changed. Until the early 1970s, for example, federal investment in energy had been substantially greater than Alberta's. This investment took the form of deductibility of resource royalties (subsequently resource allowances). Furthermore, that oil (and, subsequently, gas) requirements west of the Ottawa River had to be satisfied with western oil was a huge indirect subsidy – initially by guaranteeing demand to western producers, and subsequently by equalizing price differentials for Ontario consumers. The result is a pipeline network that runs west to east – and the Albertan battlecry "Let the Eastern bastards freeze in the dark!"

Over the past 40 years, however, there has been a paradigmatic shift whereby strategic investments in energy are now made and guided by provinces: oil in Alberta; hydroelectricity in British Columbia, Manitoba, Quebec, and Newfoundland and Labrador and, to a lesser extent, Ontario; nuclear, wind, and sun to support province restructuring to a low-carbon electricity system; the prospect (or lack thereof) of fracking across Canadian jurisdictions, and so on. Insofar as we find an energy strategy in Canada, over the past 40 years the impetus has shifted from the federal to the provincial level of government. Effectively, the "downloading" of some dimensions of environmental assessments from the federal government to the provinces reinforces this paradigm shift. Reducing federal leverage over energy bolsters the provincial purview over energy and exacerbates the sort of collective-action problems that thwart a coherent intergovernmental approach. Absent a national strategy, energy flows across provincial jurisdiction end up being guided by market forces. They follow demand and supply on the one hand and available infrastructure on the other hand.

DISCUSSION: PROSPECTS FOR FORGING A NATIONAL STRATEGY

A plethora of dynamics militate against a national energy strategy. Canada has plenty of energy, it is just not equitably distributed – and neither, in consequence, is fiscal capacity owing to differential endowments in energy revenues (Courchene 2013). Canada does not actually have an energy-resource problem per se, merely a distributional problem that it has largely relegated to the provinces to resolve for their respective consumers. Federal countries with a national energy strategy tend to be deficient in energy endowments; their energy strategy is focused on procurement. Since international trade is typically a federal responsibility, such a strategy tends to be uncontroversial. By contrast, a strategy whose main objective is distributional will necessarily be controversial, precisely because its very premise is interregional imbalances in supply and demand as a result of territorially differentiated natural-resource endowments. Canada imports some of its energy, because it has thus far proven cheaper and easier to import oil from the Middle East to Eastern Canada than to pipe it there from Western Canada. However, the economics of this calculation may be changing. So, the question really is whether Canada needs a national energy distribution strategy. Since the bulk of the necessary critical infrastructure is in private hands or with Crown corporations, that strategy, presumably, would have to rely on incentives; the federal government does not have the necessary interest, expertise, or financing to nationalize or build more critical-energy infrastructure itself. Regulation is a relatively inexpensive alternative to government investment, especially during fiscally austere times. An alternative way for government to influence the private sector is through incentives and subsidies. The provincial and federal governments both provide such incentives to the oil patch, for instance. The most glaring one is the deductibility of resource allowances, which is so huge that it impairs the federal government's capacity to finance equalization. What may appear like a bilateral arrangement between one province and the federal government actually has significant national consequences that impair intergovernmental coordination (Courchene 2013).

What Canada does need, by contrast, is a national energy export strategy. Energy exports have become a rising revenue generator for provincial and federal governments and a significant source of employment. In a fiscally constrained environment where the opportunity to hike taxes is limited, energy has become a major focal point for government to increase revenue. In 2010, for instance, Canadian exports totalled about $400 billion, of which energy made up about 20 percent, with crude oil accounting for about $52 billion and gas about $19 billion. An export strategy is needed because energy resources are often shipped internationally from a different province than the one where they were extracted, and the federal government has primary responsibility for interprovincial and international oil and gas pipelines. The same issue necessitates federal direction for an interprovincial distribution strategy: with oil and gas especially, energy tends to be consumed in provinces different from the ones where it is extracted.

The same is not true for electricity though, much of which is consumed in the province where it is produced and the remainder exported interprovincially or internationally without intermediaries. Ontario, for instance, has a total of 26 electricity interties with two provinces and three American states (Independent Electricity System Operator and the Ontario Power Authority 2014).

With flows predominantly north-south and out of the country rather than across the country, intergovernmental agreement for a national energy strategy is likely to be difficult to obtain: by virtue of extracting and/or producing different energy sources destined to different places abroad or to various American states, provincial priorities and interests with respect to energy are highly heterogeneous. Effective coordination is further complicated by asymmetry of the federal roles in oil, gas, and electricity: the federal government continues to maintain a substantial (revenue) stake in oil and gas but has largely abrogated electricity to the provinces. Moreover, maximizing the financial return on these provincial resources is a priority both for industry and for provincial governments. Because of the relatively small size of the Canadian market, and the limited amount of refining capacity, more often than not the highest bidder is found beyond Canada's boundaries. The impediment to maximizing those returns is the necessary infrastructure to get energy sources to the highest bidder – and the environmental concerns, in Canada and abroad, about the tar sands.

An interprovincial consensus on a national energy distribution strategy is conceivable, but infrastructure priorities differ among provinces, depending on whether their respective focus is on oil, gas, or electricity. Alberta's visceral reaction to Trudeau's National Energy Program and the legacy of mistrust it fostered among western provinces vis-à-vis the federal government on matters of energy is the case in point. Similarly, the legacy of the Churchill Falls electricity agreement between Newfoundland and Labrador and Quebec continues to give pause for thought to provinces looking to strike long-term bilateral energy deals (of the sort Ontario and Quebec are exploring). The stiff subnationalistic resistance Hydro-Québec ran into when it attempted to buy NB Power exemplifies just how closely energy is wrapped up in provincial identity. Interjurisdictional squabbles about new pipelines are legendary: Alberta and British Columbia, Alberta and Ontario, Canada and the United States.

Energy transcends mere rational interest. As a result, audience costs – understood as the political punishments leaders suffer for reneging on their public threats and promises – for any government looking to enter into an interprovincial agreement on energy, even merely a bilateral one, can turn out to be, as New Brunswick premier Sean Graham discovered in 2010, prohibitive. Moreover, different provinces hold different values, especially with respect to environmental protection, preservation, and sustainability, as interprovincial differences with respect to fracking illustrate.

Even if a consensus around distribution and infrastructure could be reached in principle, a consensus on its implementation may be even harder to reach. For instance, under the Kyoto Protocol with its premise on punishing producers ("Make polluters pay!") rather than consumers, further development of Alberta's tar sands will almost certainly have to be coupled to emission cuts in the rest of the country. Under the federal government's current approach to this policy field, that would mean an even greater disproportion of benefits accruing to Alberta and its producers than is already the case, while the rest of the country is saddled with a disproportionate amount of the costs associated with cutting carbon emissions in a national zero-sum game. This does not bode well for intergovernmental cooperation on energy and would require the federal government to change course on multiple fronts: resource-allowance deductions, distribution of the costs of curtailing carbon emissions, and the way and extent to which energy revenues are equalized across the country.

CONCLUSION

The fundamentals of the problem behind forging a national energy strategy are similar across federations with large territories: the United States, Australia, and India, for example. That is, their energy (re)sources are distributed unevenly, and they have just as great a distribution problem. Constitutionally, however, the federal governments in the United States and Australia have greater national powers in respect of energy than does Canada's federal government. As a result, regional differences over priorities and interests necessarily become a matter of national conflict and priority, and are largely carried out and settled in the federal political arena. As in Canada, critics in the United States and Australia regularly lament the absence of an actual national strategy which, presumably, is explicable as a function of the abundant resources in both federations. India, by contrast, whose states also enjoy considerable jurisdictional power in matters of energy, has challenges similar to Canada in forging a national strategy and implementing it – a significant impediment to India realizing its full potential for economic growth.

The asymmetry in policy approaches and substate strateties to which the constitutional division of powers in Canada with respect to energy gives rise, the way it has (and has not) been used by provincial and federal governments, and the way energy usage has changed in recent decades militate against a grand, horizontal and vertical intergovernmental bargain on energy policy and strategy. Trying to force one is bound to falter. Provincial and federal governments are thus left to forge their own energy frameworks through targeted incentives, often in the form of subsidies. As the German federation recently learned from sinking EUR 100 billion in subsidies to encourage a national strategy on renewable energy to materialize, the use of economic incentives can prove exorbitantly expensive while generating little actual return. Nonetheless, with critical infrastructure largely in private hands, and absent an intergovernmental consensus, the fallacy of composition is unlikely to be overcome, absent a national strategy.

REFERENCES

Agnew, John. 1994. "The Territorial Trap: The Geographical Assumptions of International Relations Theory." *Review of International Political Economy* 1 (1): 53-80.

Alternatives North. 2008. *A Response to the "Road to Improvement" Report by Neil McCrank on Regulatory Systems across the North*. Yellowknife. http://aged.alternativesnorth.ca/pdf/ANResponseToTheMcCrankReport.pdf.

Angell, Angela C., and John R. Parkins. 2011. "Resources Development and Aboriginal Culture in the Canadian North." *Polar Record* 47 (1): 67-79.

Bowles, Paul, and Henry Veltmeyer. 2014. *The Answer Is Still No: Voices of Pipeline Resistance*. Toronto: Brunswick Press.

Brenner, Neil, Bob Jessop, Martin Jones, and Gordon Macleod. 2003. *State/Space: A Reader*. Oxford: Blackwell Publishing.

Canadian Association of Petroleum Producers. 2014. *Crude Oil: Forecast, Markets & Transportation*. Report (June). http://www.capp.ca/forecast/Pages/default.aspx.

Canadian Electricity Association. 2015. *Canada's Electricity Industry*. http://www.electricity.ca/media/Electricity101/Electricity101.pdf

Caulfield, Richard A. 2000. "Political Economy of Renewable Resources in the Arctic." In *The Arctic: Environment, People, Policy*, edited by Mark Nuttall and Terry V. Callaghan, 485-513. Amsterdam: Harwood.

Ceglowski, Janet. 2003. "The Law of One Price: Intranational Evidence for Canada." *Canadian Journal of Economics* 36 (2): 373-400.

Courchene, Thomas J. 2013. *Surplus Recycling and the Canadian Federation: Addressing Horizontal and Vertical Fiscal Imbalances*. Fiscal Transfer Series no. 6. Toronto: Mowat Centre. http://mowatcentre.ca/pdfs/mowatResearch/87.pdf.

Engel, Charles, and John H. Rogers. 1996. "How Wide Is the Border?" *American Economic Review* 86 (5): 1741-79.

Gorodnichenko, Yuriy, and Linda Tesar. 2009. "Border Effect or Country Effect? Seattle May Not Be So Far from Vancouver after All." *American Economic Journal – Macroeconomics* 1: 219-41.

Hale, Geoffrey, and Monica Gattinger. 2010. *Borders and Bridges: Canada's Policy Relations in North America*. Don Mills, ON: Oxford University Press.

Helliwell, John F. 1998. *How Much Do National Borders Matter?* Washington, DC: Brookings Institution Press.

—. 2002. *Globalization and Well-Being*. Vancouver: UBC Press.

Helliwell, John F., and Genevieve Verdier. 2001. "Measuring Internal Trade Distances: A New Method Applied to Estimate Provincial Border Effects in Canada." *Canadian Journal of Economics* 34 (4): 1024-41.

Independent Electricity System Operator and the Ontario Power Authority. 2014. *Review of Ontario Interties*. Report prepared for the Minister of Energy. http://www.ieso.ca/Documents/IntertieReport-20141014.pdf.

Loesch, August. 1954. *The Economics of Location*. 2nd rev. ed. New Haven: Yale University Press.

McCallum, John F. 1995. "National Borders Matter: Canada-U.S. Regional Trade Patterns." *American Economic Review* 85 (5): 1024-41.

Millimet, Daniel L., and Thomas Osang. 2007. "Do State Borders Matter for U.S. Intranational Trade? The Role of History and Internal Migration." *Canadian Journal of Economics* 40 (1): 93-126.

Newman, David. 2006. "The Lines That Continue to Separate Us: Borders in Our 'Borderless' World." *Progress in Human Geography* 30 (2): 143-62.

Obstfeld, Maurice, and Kenneth Rogoff. 2001. "The Six Major Puzzles in International Macroeconomics: Is There a Common Cause?" In *NBER Macroeconomics Annual 2000*, vol. 15, edited by B.S. Bernanke and K. Rogoff, 339-412. Cambridge, MA: MIT Press.

Painter, Joe. 1995. *Politics, Geography and "Political Geography."* London: Arnold.

Preston, Jen. 2013. "Neoliberal Settler Colonialism, Canada and the Tar Sands." *Race & Class* 55 (2): 49-59.

Statistics Canada. 2013. *Electric Power and Generation – Annual (CANSIM 127-0007)*. Ottawa: Statistics Canada.

Van Hinte, Tim, Thomas I. Gunton, and J.C. Day. 2007. "Evaluation of the Assessment Process for Major Projects: A Case Study of Oil and Gas Pipelines in Canada." *Impact Assessment and Project Appraisal* 25 (2): 123-37.

Wolf, Holger C. 2000. "Intranational Home Bias in Trade." *Review of Economics and Statistics* 82 (4): 555-63.

3

TERMS-OF-TRADE CHANGES, THE DUTCH DISEASE, AND CANADIAN PROVINCIAL DISPARITY

Serge Coulombe

INTRODUCTION

In an economy opened to international trade, improvements in living standards are determined in the long run by productivity gains and terms-of-trade changes. To illustrate this, suppose the economy produces only cakes that are sold in international markets for other goods. The economy will get richer when productivity gains generate an increase in the number of manufactured cakes. Improvements in terms of trade also make the economy richer when cakes are sold for higher prices relative to other goods in international markets.

The primary purpose of this chapter is to show that the favourable evolution of terms of trade during the resource boom of 2002 to 2008 has largely shaped Canadian provincial disparity in 2012. This disparity results from the uneven spread of valuable natural resources across the territory and from provincial ownership of resources. In the second part of the chapter, I will argue that the resource boom might not have been beneficial to all Canadian provinces due to a "Dutch disease."

The relative importance of productivity gains and terms-of-trade changes depends on the degrees of openness and diversification of an economy. In a large and diversified economy such as the United States, exports do not account for a substantial portion of GDP (only 14 percent in 2011[1]), and the export base is well diversified. We should not be surprised that American economists are not very interested in measuring and analyzing the contribution of terms-of-trade changes

---

[1] Data source for export to GDP is the World Bank, http://data.worldbank.org/indicator/NE.EXP.GNFS.ZS.

to living standards. For American economists, the only driver that matters is productivity gain. Canadian economists, however, should devote more attention to terms-of-trade changes since exports in Canada account for a larger part of GDP (31 percent in 2011). Furthermore, Canada is a net exporter of natural resources, and it is well known that the prices of energy and other commodities are more volatile than for manufacturing goods in international markets. Canadian regional economists should devote even more attention to terms-of-trade changes given that Canada is a vast and sparsely populated country and its large endowment of natural resources is not evenly distributed across provinces. A whole is usually more diversified that any of its parts; consequently, most Canadian provinces, with the exception of Quebec and Ontario, are not as diversified as Canada.

In a nutshell, the smaller the economy is, the more open it is, and the less diversified it is (three things that usually go together), the more the evolution of welfare is determined by the good fortune of terms-of-trade changes. Given that the evolutions of commodity prices are much more volatile than those of manufacturing goods, the effects of terms of trade on living standards are exacerbated in a small resource-based economy.

Labour productivity, the primary concept of productivity, is measured as the ratio between the quantity of output (GDP) and the units of labour (hours worked). Productivity gains can be achieved by three sources: (1) giving labour better and more tools (capital deepening); (2) improving the skills of the labour force with education, training, and/or learning-by-doing (human capital); and (3) adopting new and more efficient technologies (multifactor productivity growth).

Canada has long been recognized in the abundant productivity literature as a very poor performer among developed economies (see for example, Tang, Rao, and Li 2010). In a previous study (Coulombe 2011), I report that productivity growth in Canada between 1990 and 2004 was on average one-half percentage point per year smaller than the average of OECD countries. In matters of productivity growth, one-half a point per year is a substantial number. Suppose that two economies, A and B, initially have the same per capita GDP and that economy A is able to sustain for a long period of time a productivity growth one-half percentage point per year larger than economy B. After 140 years, economy A will be twice as rich as economy B.

An economy can also become richer if the prices of goods it is selling to its trading partners are increasing compared with the prices of the goods it is purchasing. Changes in terms of trade may be best viewed as the relative evolution of the prices of exports and imports. For example, when the prices of manufacturing goods are slowly decreasing in world markets and the prices of commodities are increasing, as was the case between 2002 and 2008, terms of trade of resource-rich countries like Canada, Australia, and Norway, and provinces like Alberta, Saskatchewan, and Newfoundland, are improving. It is important to point out that terms-of-trade changes work both ways. When the relative price of potash for example increases in world markets, the terms of trade for potash sellers improve whereas they deteriorate for purchasers of potash.

Terms-of-trade gains might result from a favourable industrial structure when an economy, because of historical accidents or well-thought industrial policy, has acquired a know-how in producing goods and services with raising relative prices. Kohli (2004) has argued that it might be the case of Switzerland, which saw its terms of trade improve by 34 percent between 1980 and 1996. Generally speaking, however, if productivity gains result from working hard and being smarter, terms-of-trade gains result from luck (Coulombe 2011). To a large extent, natural resources are a matter of luck: the endowment is determined by geography, the ownership by law and political competition, and the commodity prices by international markets.

Luck is also very important in the case of Canadian provinces since thanks to historical events (see Plourde 2010), ownership and most of the resource revenues (including for offshore oil extraction) were granted to provinces. Anderson (2012) analyzes the division of power between local, provincial, and federal governments regarding the ownership, management, and resource revenues for petroleum in 12 federations (Argentina, Australia, Brazil, Canada, India, Malaysia, Mexico, Nigeria, Pakistan, Russia, United States, and Venezuela). He shows that for onshore resources, Canada is the only federation in which provinces not only own and manage the resource but also receive revenues from its exploitation. For offshore resources, even if the resource ownership and management are both within federal jurisdiction by the constitution, Canada is the only federation where the resource revenues have been ceded entirely to the provinces.

In the following section, I will show that terms-of-trade changes that occurred during the resource boom of 2002–2008 are the key driver of the actual provincial disparity in Canada.

LUCK MATTERS, NOT PRODUCTIVITY

Before looking at numbers, I briefly highlight the methodology used to derive the provincial and national terms-of-trade effect. The data on productivity (used for Figure 4) and terms of trade (Figures 1 and 3) were taken from Coulombe (2011). The data on labour productivity are straightforward and are measured by provincial real GDP divided by the number of hours worked.

I derived my own data on terms of trade following a simple methodology that I have developed in my studies on Canadian provincial convergence that go back to Coulombe and Lee (1995). The methodology is based on the concept that terms of trade are included in the measure of nominal GDP but are excluded from the measure of real GDP. This is why real GDP growth (and productivity growth) is a very incomplete measure of improvement in living standards for regional economies that are experiencing rapid changes in their terms of trade. For big economies such as the United States, terms-of-trade changes are not important on a year-to-year basis or in the long run, and real GDP growth is a good proxy for improvement in living standards.

To measure the changes in terms of trade, I first deflate nominal GDP (Canada and the ten provinces) using a national consumer price index (CPI).[2] With this, I get a concept analogous to national income. Terms-of-trade changes are measured as the difference between the growth of the CPI-deflated nominal GDP and the growth of real GDP. With this methodology, it is important to point out that I capture the effect of terms-of-trade changes for a Canadian province resulting from both its international and interprovincial trade. This simple methodology provides results that are generally very similar to measures obtained by more sophisticated approaches such as those employed by Kohli (2004) and Diewert and Yu (2012).

During the resource boom period between 2001 and the fall of 2008, improvements in terms of trade accounted for 30 percent of the progress of living standards in Canada. The key point that comes out of Figure 1, which depicts the provincial distribution of the windfall, is that this good fortune was not equally spread. A positive (negative) number in this figure indicates that terms of trade improved (deteriorated) on average during the 2002–2008 period.

Figure 1: Contribution of Terms-of-Trade Changes to Provincial Income Growth, Annual Percent Average 2002–2008

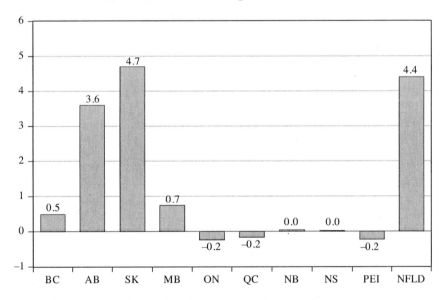

Source: Terms-of-trade data are from Coulombe (2011).

---

[2] Provincial CPI cannot be compared in levels between provinces, only in growth rates.

Saskatchewan, Newfoundland, and Alberta, three provinces that account for only 15 percent of the Canadian population, were the winners of the resource-boom lottery. In Saskatchewan, with an annual improvement in living standards generated by a terms-of-trade gain of 4.7 percent, the resource boom accounted for a 33 percent increase in national income in just six years. This is close to the 34 percent improvement in living standards that Switzerland achieved over a 16-year period as mentioned by Kohli (2004). Newfoundland and Alberta also benefited from a substantial bonanza. By contrast, the effect of terms-of-trade changes is almost null, or slightly negative, in Central Canada and the Maritimes. The terms-of-trade effect is positive in British Columbia and Manitoba but on a much smaller scale than in the three booming provinces.

I am using the fiscal capacity before equalization as the indicator of a province's living standard. The numbers for the fiscal year 2011–2012 shown in Figure 2 were obtained directly from Finance Canada. These numbers, normalized at 100 for the provincial average, are a measure of the capability of a province to raise revenues and to provide quality public services at reasonable tax rates.

The provincial disparity painted in Figure 2 is substantial. In Alberta and Newfoundland, the fiscal capacity is more than twice that of any of the Maritime provinces. If Alberta and Ontario were applying the same tax rate to their respective tax base, Alberta would be able to raise $1.80 for each dollar Ontario raises in tax

Figure 2: Fiscal Capacity before Equalization 2011–2012
Province Average = 100

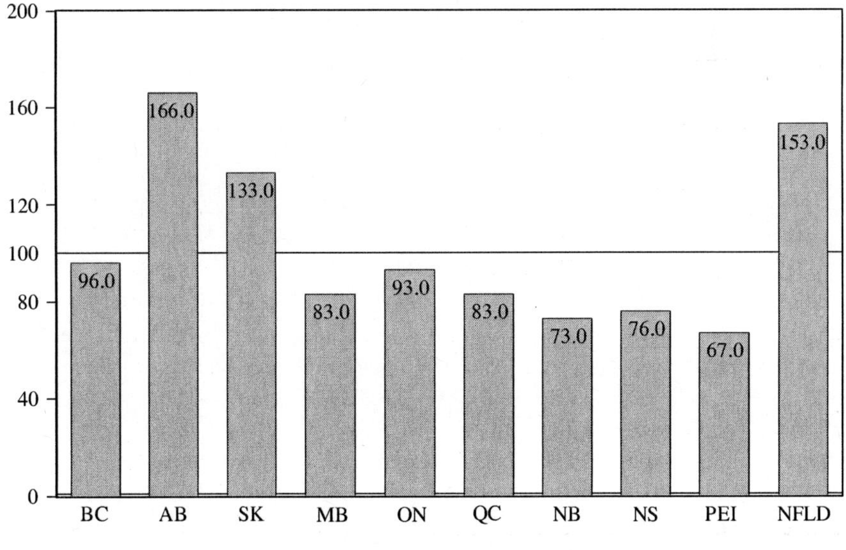

Source: Computed from Finance Canada data.

revenues. In 2011–2012, equalization was able to bring the fiscal capacity of the receiving provinces (Manitoba, Ontario, Quebec, New Brunswick, Nova Scotia, and Prince Edward Island) up to 95 percent of the province average. It is important to mention that equalization does not change the fiscal capacity of the non-receiving provinces. Consequently, even after equalization, the fiscal capacity of Alberta was 75 percent larger than for any of the receiving provinces. For Newfoundland it was 61 percent larger and for Saskatchewan 40 percent.

The striking point that comes out of the analysis of both Figures 1 and 2 is that the distribution of fiscal capacities across provinces appears to be closely related (or correlated) with the terms-of-trade changes that occurred during the resource boom period. This point is emphasized in Figure 3 with a scatter diagram.

The scatter and the fitted regression line (equation R1) illustrate the close relationship between the good fortune of a Canadian province (terms-of-trade changes during the resource boom) and living standards in 2011–2012. In the regression equation (R1), the variable *FS* (measured in 2012) stands for the fiscal capacity (Figure 2). This is our dependent variable. The variable is regressed on a constant and the terms-of-trade variable *TT* (the change between 2002 and 2008 as depicted in Figure 1). The *R*-square and the *p*-value (significance level) of the estimated coefficient (the slope in Figure 3) are shown below the regression equation.

With just ten observations (one per province) for the variables (ten points only to fit in Figure 1), it is usually extremely difficult for any variable to reach statistical significance in a regression. Statistical theory tells us that significance levels increase (*ceteris paribus*) with the number of observations. In our case, however, the coefficient of terms of trade is significant well below the 1 percent level. In order to have the slope coefficient significant at the 1 percent level with only ten observations, you need a good model, good data, and … good luck.

$$FS_{2012} = 81.3 + 15.76 * TT_{2002-2008}$$
$$R^2 = 0.83 \qquad (0.000)$$

(R1)

Another interesting result coming from the regression analysis is the high level of the *R*-square (0.83). This number indicates that 83 percent of the fiscal capacity of Canadian provinces in 2011–2012 is "explained" by the simple model that includes only a constant term and terms-of-trade changes. That does not leave much room for other explanations.

We now turn to see if productivity growth across Canadian provinces is also a significant determinant of today's portrait of Canadian disparity. The data on labour productivity during the resource boom are depicted in Figure 4. Only Newfoundland stands out in terms of labour productivity gains. With an average annual growth of 4.4 percent per year, Newfoundland clearly outpaces the other nine provinces, which average only 0.9 percent. Newfoundland's performance resulted from the

Figure 3: Luck Matters

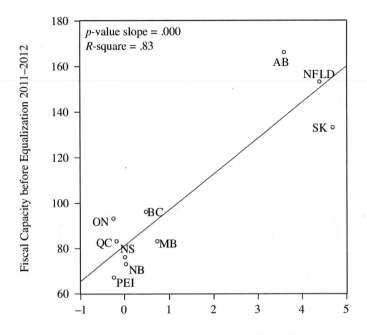

Terms-of-Trade Changes 2002–2008

Source: Terms-of-trade data are from Coulombe (2011).

shift in its industrial structure from a low-productivity-level activity, fishing, to a high-productivity-level activity, offshore oil extraction. Consequently, taking into account both the terms of trade and the productivity factors, Newfoundland stands out as the clear winner in the resource boom across Canadian provinces.

The regression equation (R2) tests if labour productivity growth (the *LPG* variable in R2), together with terms-of-trade changes during the resource boom period 2002–2008, accounts for a substantial and significant contribution to provincial disparity in 2012. The answer is no. The coefficient on the productivity variable is far from being significant with a *p*-value of 0.640. Interestingly, the coefficient on the terms-of-trade variable remains significant well below the 1 percent level.

$$FS_{2012} = 83.5 + 16.53 * TT_{2002-2008} - 2.65 * LPG_{2002-2008}$$

$$R^2 = 0.83 \qquad (0.001) \qquad (0.640)$$

(R2)

Figure 4: Labour Productivity Growth, Annual Percent Average 2002–2008

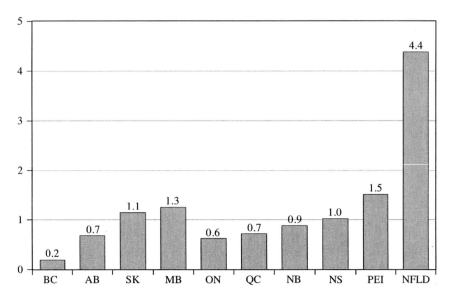

Source: Terms-of-Trade data are from Coulombe (2011).

One could argue that it takes times for productivity growth to create substantial differences in living standards such as those depicted in Figure 2 and that, consequently, the six-year study period might be too short. In the regression equation (R3), I use the mean labour productivity growth during the 25-year period from 1984 to 2009 (the whole sample used in Coulombe 2011). The results concur with those of (R2). The productivity growth variable is far from being significant. The terms-of-trade variable remains highly significant, and the small model continues to explain more than 80 percent (R-square) of the observed disparity in 2012. In equation regressions (R2) and (R3), one could also note that the estimated coefficient for the productivity variable has the wrong sign (negative). However, since the p-values in both cases are very high, these coefficients are not significantly different from 0.

$$FS_{2012} = 100.00 + 16.84 * TT_{2002-2008} - 21.3 * LPG_{1984-2009}$$
$$R^2 = 0.84 \qquad (0.000) \qquad (0.497)$$

(R3)

The results for (R2) and (R3) illustrate the robustness of the key stylized facts highlighted in this section. Provincial disparities in 2012 have been to a large extent shaped by terms-of-trade changes that occurred during the relatively short period of time between 2002 and 2008. Productivity differences across provinces do not matter statistically. Of course, economically, productivity growth does matter for Newfoundland.

The fruits from the resource boom are unevenly distributed across provinces. The consequences of these imbalances are exacerbated in Canada given the high degree of decentralization of the federation (see Boadway, Coulombe, and Tremblay 2012). Federal equalization payments transfer some of the bitumen bust to non-booming provinces, yet the fiscal capacities of the have and have-not provinces are far from being equalized at 100 percent. In the next section, I will go a step further in my argument that the resource boom has contributed to shape the current provincial disparity in Canada. I argue that the resource boom might have been detrimental to the economy of some provinces because of a Dutch disease.

REFLECTIONS ON THE DUTCH DISEASE AND PROVINCIAL DISPARITY

It has long been alleged in the development literature that an abundance of natural resources might be detrimental for economic development. Van der Ploeg (2011) provides a survey of the so-called resource curse. The Dutch disease is a more specific phenomenon that can affect the growth patterns and the industrial structure of well-developed countries. A Dutch disease might occur when a resource boom in an economy pushes the exchange rate up and crowds out trade-exposed manufacturing industries. Furthermore, the resource boom attracts scarce labour and capital from the manufacturing sector. This resource movement factor contributes to rising costs and to a loss of competitiveness of trade-exposed manufacturing industries.

In the absence of market failures and an intertemporal dimension, a market adjustment might be better viewed as a "Dutch affair," implying a temporarily costly transfer of labour and capital from the manufacturing sector to the booming resource sector. However, the Dutch affair becomes a disease when the resource boom is over; the manufacturing base is gone and will not come back. Krugman (1987) argues on theoretical grounds that specific industries that rely on learning-by-doing might be forever lost. This case is an example of a market failure known as an externality. Once the manufacturing base is gone, it is not possible to start over again because the know-how to make manufactured goods has disappeared.

Historically, the exploitation of many natural resources has been characterized by booms and busts. Sometimes the bust arises from the complete depletion of the resource (gold rush), and all that remains is a ghost town. At other times, the bust comes with technological progress that renders obsolete a specific resource-extraction process (the Guano Boom in Peru in the mid-1800s). Booms and busts are the main reason why a Dutch disease might occur from the exploitation of natural resources. It is more delicate to plan consumption and savings from an income arising from temporary windfalls than from income generated in a steady stream by economic activities in the secondary and tertiary sectors. Some economic activities from the primary sector such as agriculture and livestock also generate a relatively steady stream of income and are not subject to generating a Dutch disease.

Of course it is not possible to know today if the manufacturing activities lost in Canada due to the resource boom will come back when the boom is over. There is too much uncertainty. It is possible, however, to quantify whether some manufacturing activities in Canada have been lost because of the resource boom. This was precisely the subject of the empirical analysis by Beine, Bos, and Coulombe (2012).

Our analysis consisted of three steps. First, we divided the evolution of the Canada-US bilateral (real) exchange rate in Canadian dollars (CAD) into a Canadian and a US component. The division was based on the observation that an exchange rate is a relative price, that is, the ratio between the value of the Canadian and the US dollar in international markets. Second, we showed that the Canadian component was driven by commodity (energy and non-energy) prices whereas the US component was not. Third, we showed that employment in the trade-exposed manufacturing sector was negatively affected by the evolution of the Canadian component during the resource boom of 2002–2008.

Our results suggest that 42 percent of the appreciation of the CAD during the 2002–2008 period was due to the resource boom (the Canadian component). The remaining 58 percent was due to the depreciation of the US component following the bust of the dot.com bubble and other events unrelated to the Canadian resource sector. We also found some evidence that the appreciation of the CAD resulting from commodity prices has harmed employment in trade-exposed manufacturing industries. Specifically, 100,000 jobs lost in the manufacturing sector between 2002 and 2008 can be related to the resource boom and the consequent appreciation of the CAD (Beine, Bos, and Coulombe 2012). Not all manufacturing industries were affected, but those affected negatively were generally exposed to international competition. We used an dynamic adjustment model to calculate these employment estimates. Consequently, the numbers reported represent long-run job losses.

Other factors have also contributed to the contraction of the manufacturing sector during the resource boom. Only the lost employment associated with the Canadian component is Dutch-disease related. That accounts for only 31 percent of the 328,000 jobs lost in the manufacturing sector in Canada between 2002 and 2008. Another 180,000 jobs (55 percent) were lost due to the depreciation of the US component that occurred mainly between 2002 and 2005. Finally, 46,000 jobs (14 percent) were lost as a result of the structural decline in the manufacturing sector that has affected most developed countries, in part due to the rise of China's economy.

Since the end of 2008, the developed economies have gone through five years of financial crisis, deep recession, and Euro crisis. The US economy, still our main trading partner, has been particularly affected by the turmoil. The resource sector was negatively affected early in the crisis but has since rebounded. The manufacturing sector was also negatively affected initially due to the substantial drop in international trade. Consequently, most of the employment lost in Canadian manufacturing since the beginning of 2009 has been cyclically related and cannot be associated with a structural shift such as a Dutch disease.

Manufacturing employment in Canada is unevenly distributed. The best way to understand the geography of the Canadian economy is to borrow the core-periphery model of Krugman (1991). In this celebrated (Nobel Prize) modelling, Canada served as a practical illustration. Krugman assumes economies of scale for the production of manufacturing goods and transportation costs. He argues that the production of manufacturing goods will tend to concentrate geographically in the centre, or the core, of the economy. The main economic activity in the periphery will be based on the exploitation of natural resource and other non-traded goods. That would be the equilibrium distribution of economic activity in a closed economy. According to Krugman (1991), Canada was developed as a closed economy beginning in 1879, following the National Policy of John A. Macdonald. The policy introduced high tariffs, and as a result trade followed an artificial east-west axis where manufacturing goods were provided by the core located in southern Ontario and Quebec. The gradual opening of Canada to US trade after WWII did not alter the core-periphery structure because of scale economy. After the removal of tariffs, the Canadian core continued to be an efficient supplier of manufacturing goods and was able to remain competitive thanks to its excellent location close to two cores in the US economy: New York and the Great Lakes.

The two ingredients of my "model" of Canadian regional economies nowadays are Krugman's core-periphery model and the Dutch disease. A booming periphery with rising prices of various commodities and the acceleration of oil-sand production are detrimental to the Canadian core, which is located in the Quebec–Windsor axis. If we add to this model the negative impact of the 2008 financial crisis on world trade and manufacturing output, and the rise of China as the leading world manufacturer, Canada's manufacturing core is going through difficult times.

CONCLUSION

This chapter contributes by diagnosing rather than treating the problem. The reader mostly interested in treatment is referred to Boadway, Coulombe, and Tremblay (2012). This reader, however, should remember that a good diagnostic is often a necessary condition for successful treatment – and the avoidance of unnecessary therapy.

The problem under study is the substantial level of provincial disparity in living standards in Canada. This is a serious problem because Canada is a highly decentralized federation. Consequently, Canadians might have to move closer to the oil barrel in order to have access to good health and education services at affordable tax rates. Equalization, the constitutionally entrenched tool designed to solve this problem, can alleviate only a fraction of the disparity in provincial fiscal capacities.

I have argued here that the problem results from good fortune (terms of trade) rather than hard work (productivity). The level of disparity in the provinces' fiscal capacity in 2012 is almost solely determined by the evolution of terms of trade during the 2002–2008 resource boom. That Ontario, historically always a rich province until 2009, is now receiving equalization has nothing to do with its relatively weak productivity performance. The Canadian productivity problem is certainly a very unpleasant national problem. However, weak productivity growth is not a determinant of living standards differences across Canadian provinces at the present time. Of course, improving productivity performance in Canada will be good for Ontario and for all Canadians. But it will not cure the disparity problem.

If productivity gains are good for everyone, terms-of-trade changes generate winners and losers by definition. Newfoundland, Saskatchewan, and Alberta are the clear winners. Ontario, Quebec, and the Maritimes are the losers. They did not see the terms-of-trade improvements, but they have to live with a higher exchange rate.

Whether the problem encountered by Canadian non-booming provinces is simply a costly adjustment process (Dutch affair) or a serious problem (Dutch disease) cannot be answered now. It depends on the duration of the resource bonanza in Canada and on what happens when it eventually ends. What we do know, however, is that the necessary conditions for a Dutch disease are met in Canada. Only the future will let us know if these conditions were also sufficient. But as the wise know, nothing is harder to predict than the future.

REFERENCES

Anderson, George. 2012. "Reflections on Oil and Gas in Federal Systems." In *Oil and Gas in Federal Systems*, edited by George Anderson, 371-408. Don Mills, ON: Oxford University Press.

Beine, Michel, Charles S. Bos, and Serge Coulombe. 2012. "Does the Canadian Economy Suffer from Dutch Disease?" *Resource and Energy Economics* 34 (4): 468-92.

Boadway, Robin, Serge Coulombe, and Jean-François Tremblay. 2012. "The Dutch Disease and the Canadian Economy: Challenges for Policy-Makers." Prepared for Thinking outside the Box, a conference in celebration of Thomas J. Courchene, Queen's University, October.

Coulombe, Serge. 2011. "Lagging Behind: Productivity and the Good Fortune of Canadian Provinces." Commentary No. 331, C.D. Howe Institute, Toronto.

Coulombe, Serge, and Frank C. Lee. 1995. "Convergence across Canadian Provinces, 1961 to 1991." *Canadian Journal of Economics* 28 (4): 886-98.

Diewert, W. Erwin, and Emily Yu. 2012. "A Canadian Business Sector Data Base and New Estimates of Canadian TFP Growth." Discussion Paper 12-04, Department of Economics, University of British Columbia.

Kohli, Ulrich. 2004. "Real GDP, Real Domestic Income, and Terms-of-Trade Changes." *Journal of International Economics* 62: 83-106.

Krugman, Paul. 1987. "The Narrow Moving Band, the Dutch Disease, and the Competitive Consequences of Mrs. Thatcher: Notes on Trade in the Presence of Dynamic Scale Economies." *Journal of Development Economics* 27 (1–2): 41-55.

—. 1991. *Geography and Trade*. Cambridge, MA: MIT Press.

Plourde, André. 2010. "Oil and Gas in the Canadian Federation." Working Paper No. 10-001. Roberta Buffet Center for International and Comparative Studies, Northwestern University, Evanston, Illinois.

Tang, Jianmin, Someshwar Rao, and Min Li. 2010. "Sensitivity of Capital Stock and Multifactor Productivity Estimates to Depreciation Assumptions: A Canada-U.S. Comparison." *International Productivity Monitor* 20 (Fall): 22-47.

van der Ploeg, Frederick. 2011. "Natural Resources: Curse or Blessing?" *Journal of Economic Literature* 49 (2): 366-420.

III

Energy and Equalization

4

A NATIONAL ENERGY STRATEGY FOR CANADA: GOLDEN AGE OR GOLDEN CAGE OF ENERGY FEDERALISM?

Monica Gattinger

"If people put their differences aside and work towards a common goal and vision, results can be achieved. A truly national vision for energy that we can take to the rest of the world requires us to set our sights high. We can achieve this."

Alison Redford, Premier of Alberta, 2011–2014
Speaking to the Economic Club of Canada
Toronto, Ontario, November 16, 2011

"This report emphasizes the need for all levels of government to collaborate to create a pan-Canadian energy strategy."

The Standing Senate Committee on Energy,
the Environment and Natural Resources, July 2012

"A Canadian Energy Strategy for the 21st century is needed. One that is pan-Canadian and collaborative…. Tradition should not restrict our thinking."

Energy Policy Institute of Canada, August 2012

I would like to thank the anonymous external reviewers of this text, whose detailed and insightful comments greatly strengthened the chapter's arguments and analysis. Thanks are also due to Rafael Aguirre Ponce, Stephen Blank, John Dillon, Bruce Doern, George Hoberg, David Runnalls, and Jeffrey Simpson for their very helpful comments on an earlier draft. An energy industry expert also gave generously of his time and expertise to comment confidentially on the text. I am likewise grateful to participants at the State of the Federation 2012 Conference (November 29–December 1, 2012, Kingston, ON) for their questions and comments on my presentation, and to David Houle (Toronto) for his stimulating comments as discussant when I presented this text to the 2013 Annual Conference of the Canadian Political Science Association (June 4–6, Victoria, BC). Finally, Christopher Gunter, doctoral candidate in public administration at the University of Ottawa's School of Political Studies, provided able research assistance in support of this chapter. As always, any errors of fact or interpretation are mine alone.

For more than three decades, energy federalism in Canada has been heavily influenced – some would say cursed – by the Trudeau government's National Energy Program (NEP) of 1980. Developed in the midst of the energy crises of the time, the policy called for increased Canadian ownership and control in the energy industry, a two-price policy for energy resources with preferential pricing for Canadian consumers, restrictions on energy exports, and a host of other protective measures to enhance domestic energy security and independence from world markets. The NEP was demonized by western energy-producing provinces – particularly Alberta – as an unjustified and unjustifiable intrusion of the federal government into a domain of provincial jurisdiction, and was denounced by the United States as an attack on American energy security and US energy companies' operations in Canada. The NEP soured both federal-provincial and Canada-US energy relations and was ultimately undone, most notably by the Western Accord, which deregulated oil prices and opened the sector to international trade, and by the Canada-United States Free Trade Agreement (CUSFTA), which institutionalized free trade in energy between Canada and the United States, including explicit provisions against two-price policies and discriminatory export restrictions. Despite the NEP's demise in practical policy terms, however, it has lived on in the minds of politicians, policy-makers, and citizens alike. Provinces vigorously assert their dominance and defend their powers over energy, and they develop their respective energy policies in mostly autonomous ways. Ottawa, for its part, sticks closely to its knitting, intervening in energy or related areas in tightly circumscribed manners, knowing that provinces won't hesitate to challenge federal intervention – either in political or judicial arenas – as unconstitutional. Since the NEP, therefore, national approaches in the energy sector have been verboten, anathema to the "natural" order of energy federalism in the country, tantamount to a "third rail" in Canadian politics – say "national" and "energy" in the same breath and prepare to suffer the consequences.

Given this, growing interest and momentum in recent years for a national energy strategy represent a near conversion of things energy in the country. When touted by Alberta, the province that has most frequently played the NEP card, the change is all the more striking. What led to this shift in intergovernmental relations in the energy sphere? Does it mark a turning point to a "golden age" of energy federalism, in which governments collaborate extensively to pursue shared energy objectives? What are the contours, promise and prospects of a "national energy strategy"? This chapter seeks to answer these questions. It does so by positioning recent national energy strategy ideas in the broader historical context of intergovernmental energy relations in Canada. I develop the concept of "energy federalism," understood as the character of intergovernmental energy relations (conflictual, cooperative, collaborative, etc.), to undertake the analysis. I argue that contemporary energy federalism, which for decades has eschewed national approaches to energy and is characterized by provincial assertiveness and federal cautiousness, is but one form of intergovernmental relations in the energy sphere. Prior periods have witnessed openness to and debates over national approaches to energy, as well as greater levels of federal-provincial and interprovincial collaboration. Against this backdrop, I propose that recent national energy discussions would benefit from a more explicit focus on developing a norm of collaboration in intergovernmental energy relations, more comprehensive engagement of the Canadian public in the discussions, and greater involvement of the federal government – as partner not dominator – in the process.

The chapter is structured as follows. It begins with a brief primer on energy policy-making in the twenty-first century, a field that is increasingly complex and challenging. Governments face four demanding policy imperatives when it comes to energy: markets, environment, security, and social acceptance, what I refer to as the energy MESS. The text then develops the concept of "energy federalism" in Canada. The division of powers in the Canadian Constitution, along with the distribution of energy reserves, population, and environmental impacts of energy, has tended to produce progressively greater north-south (Canada-US) energy ties over time and, over the last three decades, intergovernmental relations tending toward competition, independence, and conflict. Given the overarching emphasis on provincial primacy, autonomy, and assertiveness in the energy sphere and the ever-present potential for hair-trigger conflict since the NEP, I characterize intergovernmental energy relations since the mid-1980s as "third rail energy federalism." As this section reveals, however, this approach represents but a recent period in Canadian intergovernmental energy relations, with prior years often characterized by greater openness to and higher levels of cooperation and collaboration among federal and provincial governments.

The next section of the chapter zeros in on the national energy strategy ideas of provincial, parliamentary, industrial, and non-governmental actors. The text reviews key proposals put forward and argues that markets, the *M* of the energy MESS, are the main driver propelling these plans. A number of fundamental changes in energy markets in North America have called into question the north-south logic of energy economics for Canada and spurred significant interest in establishing west-east and west-north energy linkages. The section explores the promise and prospects of a national energy strategy and argues that while national approaches hold promise to strengthen governments' capacities to address contemporary energy policy imperatives, intergovernmental energy relations of late suggest that Canada will not enter a golden age of collaborative energy federalism. The enduring features underpinning energy federalism, combined with the decades-long "third rail" dynamic in the sector, militate strongly against the development of comprehensive national approaches.

The chapter concludes by assessing the long-term prospects for a national energy strategy, arguing that they rest fundamentally on how the concept is defined and, most importantly, the capacity of governments to think beyond the "golden cage" of third rail energy federalism to more collaborative forms of intergovernmental energy relations. I propose that governments should begin by developing a framework agreement on energy collaboration that would identify the rationales, principles, and opportunities for collaboration and seek to build the norm of intergovernmental collaboration on energy. They should then strike an "energy deal" between the federal and provincial governments that would pursue the market access objectives of energy-producing provinces, while addressing the environmental, social and economic concerns of other provinces, Aboriginal communities, environmental groups, and ordinary Canadians.

ENERGY POLICY IN THE TWENTY-FIRST CENTURY: MAKING A MESS OF THINGS

Energy policy-making has become ever-more challenging over time.[1] It comprises four key policy imperatives that have layered progressively over one another. First, in the 1970s and 1980s, the policy focus of most western industrialized countries was on getting energy markets to work more efficiently and competitively, largely through deregulation and privatization. In Canada, this process got underway in the 1980s: in the oil and gas sector, it included deregulating prices, introducing competition into various segments of the upstream and downstream markets, liberalizing trade, and unbundling various functions within energy firms to establish open, non-discriminatory access to their services and facilities[2] (see Plourde 2005). The electricity sector followed in the footsteps of oil and gas in the 1990s, with greater competition introduced into those segments of the industry (generation and wholesale/retail supply) that could be operated under non-monopoly conditions (Plourde 2005). This first component of the energy MESS also includes energy markets in the sense of overall markets for energy in Canada, North America, and abroad: the ever-shifting dynamics of energy supply, demand, and pricing, and what they mean for Canadian energy producers and consumers in terms of market context and export opportunities.

Second, in the 1980s and 1990s, environmental considerations came increasingly to accompany the policy focus on energy qua energy markets. Mounting concerns over the environmental impact of energy exploration, production, transmission, and consumption generated new and intensified policy attention to such matters as biodiversity, ecosystem health, climate change, land use, and water quality and diversion. Given the transboundary nature of environmental matters, many of these issues have been the subject of international agreements (e.g., the Canada-US Air Quality Agreement of 1991, the United Nations Convention on Biological Diversity in 1993, and the United Nations Framework Convention on Climate Change in 1994, followed by the Kyoto Protocol of 1997 and successor agreements). In Canada, individual provinces have pursued environmental policies in relation to energy.

---

[1] For a more elaborate discussion of the energy policy MESS, see Gattinger (2012).

[2] Privatization of Petro-Canada was only to follow in the 1990s.

These policies include, for example, Alberta's levies on large emitters in 2007, Quebec's carbon charge in 2007, British Columbia's carbon tax in 2008, Ontario's Green Energy Act of 2009, and provincial participation in the US-based Western Climate Initiative, including Quebec's 2014 cap-and-trade program with California. The environment also formed the basis of a domestic intergovernmental agreement in 1998, the Canada-wide Accord on Environmental Harmonization, which lays out the objectives and principles underpinning governments' collaboration on pan-Canadian environmental issues.[3] Action on the energy-environment interface at the federal level has often focused on international treaties (e.g., UNFCCC and Kyoto), with limited action domestically beyond subsidies and voluntary measures to pursue the country's international climate change commitments (Jaccard and Rivers 2007). In recent years, the federal government has adopted a number of American policies (e.g., tailpipe emissions and US commitments at Copenhagen) and made some progress on regulations for coal-fired generation in the electricity sector.

Third, energy security concerns, while always an undercurrent of energy policy, came to both broaden and deepen at the turn of the century. In the United States, mounting reliance on foreign energy imports in the 1980s and 1990s, particularly for oil, prompted growing concern over the country's energy security. While Canada does not face these challenges to the same degree given its status as a net energy exporter, Canadian consumers and the Canadian economy are nonetheless vulnerable to energy price spikes and volatility, and eastern regions of the country are dependent on foreign oil imports for their supply. But energy security concerns of this sort apply not only to oil but to natural gas and electricity as well. Overall, energy security in Canada tends to be understood in terms of security for Canadian consumers and the economy (affordability, reliability, and security of supply).[4] Following the terrorist attacks of 9/11, the concept of energy security has broadened in scope, not only in the United States but also in Canada. The attacks focused attention on the security of critical energy infrastructure, both the physical and cyber-security of pipelines, nuclear facilities, refineries, and so on. Indeed, in the years following 9/11, Canadian energy facilities were identified by al-Qaeda as potential targets to disrupt American energy supply (Johnston 2008). Added to this were a number of smaller-scale bombings of energy infrastructure in Alberta and British Columbia, which generated increased policy attention to critical energy infrastructure protection. The concept of energy security also broadened following the "great blackout" of 2003, the largest power outage in North American history, which saw some 50 million Canadians and Americans lose power in the US Midwest, US Northeast, and Ontario. In conjunction with the progressive application of information and

---

[3] Quebec did not sign on to the agreement.

[4] The concept of energy security used here does not include security of access to energy markets for energy producers. This is captured in the "markets" component of the energy MESS, as noted in the preceding paragraphs.

communications technologies to the electricity grid (the so-called smart grid), both physical and cyber-security in the electricity sector have likewise received heightened attention. The focus on critical energy infrastructure protection has also sharpened following revelations of sophisticated and systematic hacking efforts targeting energy firms and critical energy infrastructure in Canada and the United States, allegedly by the Chinese military (Sanger, Barboza, and Perlroth 2013).

Fourth, energy policy-makers have increasingly to attend to social acceptance – or lack thereof – for energy exploration, production, distribution, and use. Energy policy and regulation used mostly to "hum along" under the political radar, but over the last number of years, it is scarcely possible to open leading dailies or listen to the news without coverage of one or more stories of public opposition to energy projects of various descriptions. Not only has public opposition intensified, it has also grown considerably in scope. Opposition in the 1980s and 1990s could pre-dominantly be characterized as NIMBYism ("not in my backyard"), but in recent times, this has progressed to far more challenging forms of principled opposition, captured neatly by the acronyms BANANA ("build absolutely nothing anywhere near anything") and NOPE ("not on planet earth"). These forms of opposition cannot always – indeed can rarely – be addressed by conventional responses in regulatory and industrial toolkits (compensating affected parties, project relocation, etc.).

Taken together, these four policy imperatives – market, environment, security, and social acceptance – constitute the complex, multifaceted policy terrain facing energy policy-makers in the twenty-first century. The question for policy-makers is, What kind of MESS will they make of energy policy: a mess in the sense of disorder and disarray (uncoordinated, ill-conceived policies) or a mess in the sense of a "mess hall," a place where people come together to meet their shared needs (policy that identifies balance-points between market, environment, and security imperatives that garner social acceptance)? The chapter returns to this question when discussing the profile, promise, pitfalls, and prospects of a national energy strategy for Canada.

ENERGY FEDERALISM IN CANADA: FROM COOPERATION AND COLLABORATION TO COMPETITION, CONFLICT, AND INDEPENDENCE

Despite energy's pivotal role in Canada – as an industry comprising a substantial proportion of the domestic economy and exports, as a motor for the manufacturing sector and economic competitiveness, as the fuel source to conquer the cold of Canadian winters and meet the transportation needs of a vast and sparsely populated country, as the source of sometimes explosive federal-provincial conflict, and as a vehicle for province-building in fiscal and cultural terms – there has been relatively limited attention to energy in recent literature on Canadian federalism. A number of seminal books on energy in Canada were published in the 1970s and 1980s (e.g., Doern and Toner 1985; McDougall 1982; Pratt 1976), and a number of volumes on Canadian federalism in the 1970s/80s addressed energy (e.g., Milne 1986; Panitch 1977; Richards and Pratt 1979), but more recent scholarship on Canadian federalism has not tended to address energy qua energy.[5] Rather, energy has been treated mainly as a subset of or in connection to the environment, particularly climate change. Here, there is much more scholarship. Scholars with dedicated research programs have systematically studied intergovernmental relations in the environmental sphere – including, notably, development of the term "environmental federalism" (see, for example, Courchene and Allan 2010).

Given this scholarly context, this chapter focuses on energy qua energy and develops the concept of "energy federalism" as a means of exploring intergovernmental relations in the sphere. Energy federalism is understood here in the simple sense of the *character* or *dynamics* of federal-provincial and interprovincial relations in the energy field. I argue that these relations are influenced by the institutional, political, geological, economic, and demographic characteristics of Canada and of the energy sector in particular. Some of these features are enduring and slow to change, and are largely constants of energy federalism: the division of powers in the Constitution and the overall distribution of economic activity and population across the country. Other factors, like market conditions in the energy sector (prices, demand/supply, etc.), politics, and energy reserves, can change rapidly.[6] The combination of these factors and their various configurations over time shape the character of intergovernmental energy relations. To flesh this out, I begin by sketching out the factors. I then develop a spectrum characterizing different forms of intergovernmental policy relations to explore the changing nature of federal-provincial and interprovincial energy policy relations over time.[7]

Energy federalism is shaped first and foremost by the constitutional division of powers in the field. It has been said that Canada has one of the most divided and decentralized constitutional arrangements for energy among western industrialized countries (Doern and Gattinger 2003). Exploring these arrangements through the energy MESS framework developed above, however, reveals that decentralization is predominantly the case when it comes to energy markets, but less so for environment and security imperatives, as well as for social acceptance, particularly when it involves Aboriginal peoples.[8] Indeed, the more that these considerations – especially

---

[5] A notable exception is Anderson (2012).

[6] In the case of reserves, changes come either from new discoveries or from the development of new technologies enabling the economic production of known reserves.

[7] This analysis is admittedly preliminary, and will require elaboration and refinement in subsequent research.

[8] Space limitations preclude a more comprehensive treatment of Aboriginal issues as they relate to energy development. Overall, Aboriginal peoples have become increasingly active

environmental – have become fundamental to energy policy-making, the more the federal and provincial governments are both central players in the energy field, and the more multifaceted, multidimensional, and complex energy federalism has become. Nonetheless, over the last number of decades, energy policy-makers have tended to approach energy federalism in an active/assertive (provinces) or passive/ cautious (federal government) way, a point to which I return below.

When it comes to energy markets, the provinces are dominant players. They have constitutional jurisdiction over non-renewable natural resources, including exploration, development, management, royalties, and intraprovincial energy trade and commerce. They also have jurisdiction over the generation, production, transmission, and sales of electricity within their boundaries (nuclear is an exception, as discussed below). The federal government's powers most closely related to energy markets derive from its jurisdiction over interprovincial and international trade and commerce (including foreign investment), international treaty-making, taxation, fisheries, and energy development offshore and on frontier lands. It bears mentioning, though, that the federal government has truncated or devolved a number of its powers in these areas: with respect to offshore and frontier lands, it has negotiated agreements with provincial and territorial governments to delegate or co-manage regulatory authority and royalties (e.g., the Canada-Newfoundland Atlantic Accord and the Yukon Territory Agreement). And with the negotiation of the Canada-US Free Trade Agreement, it in effect used federal treaty-making powers to liberalize international energy trade, thereby reducing its control over international energy flows. Since the National Energy Program, Ottawa has also been hesitant to intervene in matters of interprovincial energy trade, limiting its role mostly to the regulatory review of energy infrastructure crossing provincial boundaries. Ottawa also intervenes in the energy sector via the federal spending power and equalization. With respect to the former, recent examples include federal loan guarantees for the Muskrat Falls hydroelectric development in Newfoundland and Labrador and federal investments in carbon capture and storage projects in Alberta and Saskatchewan. With respect to the latter, the federal government decides whether or not (or under what circumstances and how) energy royalties or other provincial revenues in the energy sector are included in equalization formulas. This has generated heated debates with oil- and gas-producing provinces, which have perceived the program to unfairly reduce the equalization payments they receive during times of high energy prices (see, for example, the discussion in Courchene 2004 of "confiscatory equalization"). Equalization has also spurred significant

---

and sophisticated when it comes to energy, and through multiple court challenges and decisions have gained a number of legal rights in the realm of energy development. These include governments' "duty to consult and accommodate" Aboriginal communities on projects that might adversely affect current or potential Aboriginal or treaty rights.

conflict between Ottawa and the provinces of Nova Scotia and Newfoundland and Labrador over the status of their offshore energy royalties in the program, with both provinces securing protection for their resource revenues from "claw backs" through equalization. In broader terms, the treatment of provincial energy revenues in equalization can indirectly influence the energy sector by shaping the fiscal incentives for provinces to develop their energy resources. The federal government also plays a strong role when it comes to the development of energy resources on or crossing Aboriginal lands given its jurisdiction over reserves and in instances where it negotiates land claims or other agreements (provinces, of course, can also be key actors in these arrangements).

The environmental imperative of energy policy arguably generates the greatest level of involvement of both provincial and federal governments in the energy sphere. Provinces have jurisdiction over the conservation of energy resources within their boundaries as well as intraprovincial environmental impacts of energy. The federal government has jurisdiction over transboundary environmental impacts, as well as fisheries, navigation and shipping, agriculture, criminal law, and the power to legislate for peace, order and good government. Any single energy project is very likely, therefore, to trigger federal and provincial governments' involvement through their respective environmental powers. Recently, however, the federal government has lessened its environmental role in energy by significantly reducing the number and range of projects requiring federal environmental assessment.

The federal government continues to retain a key role in the security dimension of the energy MESS, however, through its role in critical energy infrastructure protection and in nuclear safety, the latter via the Canadian Nuclear Safety Commission.[9]

Energy federalism is also shaped by the distribution of reserves and energy production in the country (along with the technologies and infrastructure available to develop them), and the distribution of population and greenhouse gas (GHG) emissions. As shown in Table 1, established reserves, production, population, and GHG emissions are variable and regionally concentrated throughout Canada, with the province of Alberta the dominant reserve-holder and producer of oil, followed by Saskatchewan and East Coast offshore (mainly Newfoundland and Labrador but also Nova Scotia). The largest natural gas reserves are found in Alberta, British Columbia, and Saskatchewan; these provinces are also the major producers, followed by East Coast offshore (Newfoundland and Labrador and Nova Scotia). As discussed below, Canada also has vast reserves of unconventional natural gas (shale gas), but given environmental concerns over the hydraulic fracturing process ("fracking") used to develop this resource, development of shale gas is only ongoing in British Columbia and Alberta (Quebec has placed a moratorium on shale

---

[9] The federal government has also promoted nuclear industry development through the Crown corporation Atomic Energy of Canada Limited, portions of which have been privatized in recent years.

Table 1: Canadian Population, Economic Activity, Energy Reserves/Production, and Greenhouse Gas Emissions

Province /Territory	Population (thousands, 2012)	GDP (billions of dollars, 2012)	Oil (Gas) Reserves (MMbl, 2010) (Tcf, 2010)	Oil (Gas) Production (MMbl, 2012) (Tcf, 2012)	Electricity Generation (TWh, 2012)	GHG Emissions (megatonnes CO$_2$ equivalent, 2011)	GHG Emissions per capita (kt CO$_2$ equivalent per person, 2011)
British Columbia	4,622.6	217.7	117.4 (27.8)	7.7 (1.5)	69.5	59.1	12.9
Alberta	3,873.7	295.3	170,126.8 (38.8)	841.9 (4.3)	66.1	242.4	64.2
Saskatchewan	1,080.0	74.7	1,156.4 (2.3)	171.9 (0.2)	22.0	72.7	68.7
Manitoba	1,267.0	55.9	48.5 (-)	17.6 (-)	32.6	19.5	15.6
Ontario	13,505.9	654.6	9.9 (0.7)	0.5 (0.007)	140.9	170.6	12.8
Quebec	8,054.8	345.8	- (-)	- (-)	199.7	80.0	10.0
New Brunswick	756.0	32.2	- (0.1)		8.7	18.6	24.6
Nova Scotia	948.7	37.0	East Coast Offshore 887.3 (0.3)	East Coast Offshore 72.2 (0.08)	10.1	20.4	21.5
Prince Edward Island	146.1	5.4			0.2	2.2	15.1
Newfoundland & Labrador	512.7	33.6			42.3	9.4	18.3
Yukon	36.1	2.7			0.5	0.4	11.3
Northwest Territories	43.3	4.8	408.3 (0.5)	4.8 (0.007)	0.6	1.6	20.6
Nunavut	33.7	2.0			0.2		
TOTAL	**34,880.5**	**1,762.4**	**172,754.6 (70.4)**	**1,116.6 (6.0)**	**594.9**	**702**	**20.4**

Notes: Totals may not add due to rounding or methodological approaches used in these sources to calculate national figures. GDP figures are expenditure-based at current prices. Oil = crude oil and non-conventional oil. Reserves = remaining established reserves. Oil reserves for the territories include the Mackenzie/Beaufort. Natural gas reserves for the territories only include mainland reserves. Data for Alberta non-conventional oil reserves from 2011. MMbl = million barrels. Tcf = trillion cubic feet. TWh = terawatt hours.

Sources: Canadian Association of Petroleum Producers (2013); Canadian Electricity Association (2013); Environment Canada (2014); Statistics Canada (2013, 2014).

development and there is fierce opposition to shale gas in Nova Scotia and New Brunswick). In the electricity sector, all provinces generate electricity for domestic consumption,[10] but they do so with varying generation sources (hydroelectricity, coal, nuclear, natural gas, etc.). As the table reveals, energy reserves and production tend to be at a distance from major population centres; reserves are predominantly in the west, north, and east while major population concentrations are in the central provinces. This characteristic accentuates the differences in provincial GHG emissions, with major hydrocarbon producers emitting the highest volumes of GHGs either in absolute (Alberta) or per capita (Saskatchewan) terms.

The demographic context and the location of energy reserves and production have tended to produce north-south energy flows: the closest major population centres to which western energy producers ship their products have tended to be in the United States. This north-south orientation of energy markets has also been entrenched in policy with, notably, the 1961 National Oil Policy enacted by the Diefenbaker government and discussed below. In the electricity sector, provinces that export power also do so primarily in a north-south orientation. Although the 1960s, 1970s, and 1980s saw periodic discussion between the provinces and Ottawa over developing a national power grid, such talks were ultimately unsuccessful owing largely to competitive political and economic dynamics between provinces (Froschauer 1999). As such, electricity flows beyond provincial boundaries are predominantly south to US markets, with British Columbia, Manitoba, Ontario, Quebec, and New Brunswick the main exporting provinces.

Energy federalism, particularly in the wake of the National Energy Program, has been characterized by provinces developing their respective energy policy frameworks – whether for electricity, oil, or gas – in relatively independent ways and with little regard to the policies of other governments, federal or provincial. As noted below, where there are direct interactions between provinces and/or between provinces and the federal government, they often tend toward conflict (political and judicial) over interpretation of federal and provincial jurisdiction in the energy field. Ironically, at a time when scholarship on Canadian federalism has traced the shift from classical to cooperative, competitive, constitutional, and more recently, collaborative federalism (see Simeon 2010), energy federalism seems to be stuck in a competitive or classical groove in both federal-provincial and interprovincial terms.

To explore this in greater detail, Figure 1 shows a spectrum of various policy relations between governments (either federal or provincial): from relations rooted in *conflict*, where interests diverge and there is open discord, to *independent* policies, where governments develop their policies without regard for their potential consequences on counterparts (or vice-versa), to *harmonization*, where governments consciously develop common policy approaches. In between independence

---

[10] The exception is Prince Edward Island, which imports most of its electricity from New Brunswick.

and harmonization lies *parallelism*, which refers to governments adopting similar policy approaches to other governments but tailoring them to local circumstances, *coordination*, whereby governments consciously work to reduce spillovers or maximize compatibility between their policies, and *collaboration*, in which they work together to pursue common objectives.

Figure 1: Spectrum of Intergovernmental Policy Relations and Approaches to Canadian Federalism

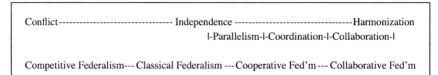

Note: This spectrum was originally developed to analyze Canada's international policy relations. It first appeared in Gattinger (2005) and was subsequently revised in a very fruitful collaboration with Geoffrey Hale (see Gattinger and Hale 2010).

As shown in the figure, these various policy relations map on to different forms of Canadian federalism identified in the literature. Various forms of federalism are often used in a temporal sense to characterize historical periods in federal-provincial relations in toto (see, for example, Simeon 2010), but I conceive of them here as differing kinds of relations between and among federal and provincial governments in individual policy sectors. As such, the spectrum developed here does not present them temporally, but rather, according to the nature or character of intergovernmental relations. Seen through this lens, competitive federalism, associated with the onset in the 1960s of federal-provincial conflict over constitutional roles and responsibilities with the key player Quebec seeking greater powers and decentralization in the Canadian federation, can be conceived of as a more general species of intergovernmental relations characterized by conflict between Ottawa and the provinces or between the provinces themselves. Classical federalism, meanwhile, associated with the pre-depression, pre-welfare-state period during which the powers of the federal and provincial governments were more balanced and there was less conflict between governments, is akin to policy relations of independence, where governments focus on working within their own spheres of constitutional jurisdiction. Cooperative federalism, linked to the postwar development of the welfare state up to the 1960s, led by the federal government with provincial participation, is situated in between independence and harmonization, given its close relationship to ideas of parallelism and coordination. Collaborative federalism, with its focus on relationships between governments as equals working toward shared goals, falls to the right of the policy relations spectrum, which features coordination, collaboration, and harmonization.

Four Broad Periods of Energy Federalism

Using the analytical framework developed above, four broad periods of energy federalism can be identified since Confederation.[11] Each places different emphasis on the various imperatives of the energy MESS, is characterized by different configurations of factors (institutional, economic, demographic, etc.), and can be positioned at a different location on the spectrum in Figure 1.

Nationalist cooperative energy federalism (1867–1930): This first period runs from Confederation to 1930, prior to the discovery and development of substantial oil and gas reserves in the country (in Turner Valley, Leduc, and Redwater, Alberta, in 1930, 1947, and 1948, respectively). This period also precedes the Natural Resources Act of 1930, which conferred natural resource rights on the Prairie provinces, whose governments (unlike other provincial governments) were not granted jurisdiction over energy when they entered Confederation. During this period, coal was the dominant fuel source and large quantities were being produced in Eastern Canada. Oil and gas were produced, but in such small quantities that there were no exports and production was consumed in the province in which it was produced (Plourde 2012). Given this, the country imported most of its oil and gas from the United States. While there was substantial debate between Ottawa and the Prairie provinces over control of their natural resources,[12] national approaches to energy were "on the table" and actively discussed. The federal government was a lead player in these processes, which focused on coal given its dominant role in the economy and in Canadian energy production. The debates over a national coal policy centred on national unity considerations (binding the country via energy ties) and energy security concerns (ensuring that coal produced in Eastern Canada would support the energy needs of consumers elsewhere in the country; McDougall 1982). Overall, the focus of policies was predominantly on the *M* and first *S* of the energy MESS: markets and security. Environmental considerations were yet to emerge as central issues on the energy policy stage, and social acceptance imperatives were also much less prominent.[13] Given these characteristics, this period can be termed one of *nationalist cooperative energy federalism:* national energy policies were

---

[11] The factual information in this section draws primarily on Doern and Toner (1985), Froschauer (1999), and McDougall (1982). The depiction and description of the periods is my own and is admittedly brief. Each period will need to be refined and elaborated in subsequent research.

[12] For a detailed account of this issue, see Janigan (2012).

[13] This is not to say that social acceptance played no role whatsoever. There was significant contention in Ontario, for example, around early hydroelectric development in the Niagara region, which mainly supplied power to firms on the American side of the border rather than to Ontario industry and citizens. Nonetheless, the scope, frequency, and intensity of public opposition to individual energy projects were far lesser. Indeed, major energy developments

politically acceptable to debate and undertake, the federal government played a relatively strong role in the energy field via the national coal policy, and provincial governments with resource rights developed their internal energy policies – for oil and gas, rather independently, and for coal, with an eye to accessing markets outside their boundaries. The electricity sector was in private hands with the exception of Ontario, which nationalized its system at the turn of the century.

Expansionist collaborative energy federalism (1930–1950s): The second period, which runs from 1930 to the end of the 1950s, coincides with the discovery and large-scale development of oil and gas reserves in Alberta, Saskatchewan, and British Columbia, and the extension of natural resource rights to all provinces. By the end of this period, oil and gas had supplanted coal as the dominant fuel sources. These changes saw growing provincial intervention and independence in the energy domain, including the expansion of provincial regulation to manage energy resource development. The period also saw continued intervention of the federal government to address issues of energy security and components of energy deemed to be "in the national interest." These interventions included the 1946 Atomic Energy Control Act, which established federal jurisdiction over nuclear energy, and the creation of Atomic Energy of Canada Limited in 1952 to develop and market the country's CANDU reactors.

Given the growing energy production in the country (oil, gas, and electricity), national policies and approaches to energy focused less on Canadian energy self-sufficiency and more on growth of the industry via infrastructure development and exports. The federal government enacted the Pipelines Act in 1949, which conferred it authority over interprovincial and international oil and gas pipelines. Ottawa also created the National Energy Board (NEB) in 1959 to regulate interprovincial and international energy flows and infrastructure. This was a recommendation of the 1957 Royal Commission on Energy (the Borden Commission), which analyzed the interplay between domestic and international energy markets. In addition to recommending creation of the NEB, the commission advised the federal government to divide Canada into two markets for oil, a recommendation concretized in the 1961 National Oil Policy. Henceforth, oil from western producers would meet Canadian demand in the west and as far as the Ottawa Valley, and imports would meet demand in markets east of this point, in what came to be known as the Borden Line (see McDougall 1982). In the electricity sector, Ottawa enacted the 1963 National Power Policy, which supported power development via exports to the United States.

In keeping with the first period in energy federalism, one of the characteristics of this subsequent stage was the political interest in and openness to exploring and enacting national approaches to energy on the part of both federal and provincial governments. This is not to say that discussions were not contentious – they

---

were often greeted positively by citizens eager to avail themselves of energy for lighting, heating, transportation, and the like.

were often very heated – but that these discussions were politically acceptable to undertake and it was seen to be in the national interest to do so. The National Oil Policy was one expression of this, as were interprovincial discussions facilitated by Ottawa beginning in the 1950s over the development of a national electricity grid. As noted previously, these discussions were ultimately unsuccessful, and the federal government enacted the National Power Policy in 1963 to support power exports (see Froschauer 1999, Chapter 2, and below). Nonetheless, the dynamics of third rail energy federalism characteristic of the post-NEP period were yet to emerge. Instead, there was more in the way of collaborative federalism (or at least attempts to collaborate) in the energy sphere: federal and provincial governments related more as equals in this second period than they did in the first. Prairie provinces attained their natural resource rights, and intergovernmental relations were based more on negotiation than on hierarchically driven federal leadership. Indeed, when it came to oil and gas, Doern and Toner (1985) note that the 1940s to the early 1970s were "marked by a reasonable consensus of values between the federal and provincial governments over the management of Canada's growing oil and gas reserves. The overriding objective of energy policy was to encourage oil and gas production and to stimulate the growth of the domestic petroleum industry" (67). But this period of *expansionist collaborative energy federalism*, which was focused predominantly on developing energy markets (the *M* of the MESS framework), was not to last.

Competitive energy federalism (1960s–mid-1980s). The 1960s began to see greater conflict between the federal and provincial governments over energy, with provinces becoming more assertive and staking and reinforcing their constitutional powers in the energy sphere. Provincial coffers in Alberta, British Columbia, and Saskatchewan were also increasingly benefiting from energy royalty revenues, and electricity-exporting provinces were reaping the rewards of sales to the United States.

The nationalization of electricity in Quebec, emblematic of the *maître chez nous* dictum of the Quiet Revolution, saw a much more self-interested dynamic on the part of the province: electricity and Hydro-Québec became intimately tied with cultural identity and provincial autonomy. Indeed, the Quiet Revolution was one of the main reasons ideas for a national power grid advanced by Prime Minister Diefenbaker in the early 1960s fell apart. The project was reoriented from national to continental by his successor, Lester Pearson, with the 1963 National Power Policy. Other jurisdictions also completed their shift to public electricity systems during this period (British Columbia and Manitoba in 1961; Newfoundland in 1974) and, along with Ontario, which had had a public system since 1906, put the focus on developing their electricity sectors with large infrastructure builds in the 1960s and 1970s.

These included, of course, the Churchill Falls project in Newfoundland and Labrador. This massive hydro development is a particularly sour point for Newfoundland and Labrador in its relations with Quebec and the federal government. When the project was in development in the 1960s, Newfoundland wanted the Pearson government to pass legislation to establish a power corridor through Quebec so that it could sell electricity directly into the US market, but Pearson was concerned that if he intervened, the Liberal Party would lose support in Quebec. Without support from Ottawa, Newfoundland was left in the position of negotiating with Hydro-Québec, the sole potential buyer. The two provinces signed an agreement that in hindsight was extremely disadvantageous to Newfoundland and Labrador. The province agreed to sell power to Quebec at a very low rate locked in for 65 years. As of the mid-2000s, Quebec was benefiting from this agreement to the tune of about $850 million a year (Dunn 2005). Ottawa's decision not to intervene left Newfoundland and Labrador mistrustful of the federal government and bitter in its relations with Quebec.

But conflict over energy in this period was far from confined to electricity. When the energy crises of the 1970s hit and world prices for oil and gas soared, the stage was set for explosive conflict between Ottawa and the western-producing provinces and between Eastern and Western Canada. As Doern and Toner (1985) note, post-1973, "government-industry and inter-governmental political conflict became much more prevalent" (67). The federal government sought to protect eastern consumers from world prices and enhance Canadian energy security with domestic production. Western producers went along up to a point, but as Ottawa's demands began to cut further and further into profits and royalties, western tolerance withered. Tensions ran particularly high between Ontario and Alberta, with Alberta even threatening to cut off gas exports to the east, and Ontario announcing in 1973 that it would test the constitutionality of the threat. As the energy crisis deepened, Ottawa, intent on shielding Eastern Canada from world oil prices, brought in the National Energy Program in 1980. This policy was a bridge too far for western-producing provinces, which characterized it as unwarranted federal intervention into a sphere of provincial competence, given its direct impact on provincial capacities to develop, manage, and raise revenue from energy resources. Alberta successfully challenged the constitutionality of the federal government's export tax on natural gas (Chalifour 2010, 179). Given these dynamics, this period, which focused mainly on the *M* and first *S* of the energy MESS (markets and security), can be termed one of *competitive energy federalism* – not only between Ottawa and energy-producing provinces, but also between producing provinces in the west and consuming provinces in the east.[14] The period ended in the mid-1980s when the federal government rolled back the last of the NEP measures and negotiated the Canada-US Free Trade Agreement (CUSFTA), which institutionalized free trade in energy. The period also included constitutional amendments in 1982 granting provinces the power to levy indirect taxes on their natural resources.

---

[14] The popular bumper sticker of the time in Alberta, "Let the eastern bastards freeze in the dark," vividly captures this point.

Third rail energy federalism (mid-1980s–): The fourth period begins in the mid-1980s after the conflict over the National Energy Program. While one might have expected that the demise of the NEP and the coming into force of CUSFTA would bring peace to the energy federalism landscape and set the stage for collaboration and cooperation, what has occurred instead is entrenchment of a norm of provincial assertiveness and federal reticence when it comes to intergovernmental energy relations. Provinces vigorously assert their dominance in the energy sphere and develop their respective energy policies relatively independently from one another, while the federal government treads lightly in the energy field, ever wary of raising provincial hackles. Ottawa has even circumscribed its jurisdiction in energy, not only through CUSFTA, but also through such measures as the Atlantic Accords with Nova Scotia and Newfoundland and Labrador. These agreements, which enable the two provinces to tax offshore revenues as if they were the resource owners and share regulatory authority with the federal government through joint regulatory boards, have led to a situation where "it is clear that the major decisions regarding offshore activities are effectively controlled by the provincial governments" (Plourde 2012, 101). Plourde (2012) sums up the situation nicely, saying that when it comes to the oil and gas sector, since the mid-1980s "the federal government has disengaged from active intervention in this area, relying instead on market forces to influence developments and effectively adopting a rather broad interpretation of provinces' constitutional authority to control the exploitation of natural resources located within their boundaries" (94-95).

This is not to say that there has been a complete absence of federal-provincial and interprovincial collaboration (e.g., the federal government has been accommodative, if not outright supportive, of provinces' efforts to develop and export their energy resources), but rather that there has been a constant undercurrent of federal cautiousness and provincial assertiveness in intergovernmental energy relations. This, along with the decades-long political prohibition on national approaches to energy, characterizes this fourth period as *third rail energy federalism*: get too close to provinces' energy powers or dare to propose national approaches to energy and suffer the consequences.

This dynamic has been particularly evident as environmental considerations have ascended on the energy policy agenda. Relations between Ottawa and the provinces can be heated, with the latter arguing – either politically or through the courts – that federal policies must not hamper provincial capacities to develop energy resources. Through the courts, this includes Ontario challenging federal jurisdiction in the nuclear sphere in 1993 and Hydro-Québec taking the federal government to court in 1997 over the constitutionality of federal environmental protection measures (Chalifour 2010, 180 and 184). But it is in the political realm where these dynamics are most frequently on display. One of the more memorable moments took place in 2002, when Alberta premier Ralph Klein spoke out publicly and unexpectedly against Prime Minister Jean Chrétien's Kyoto targets for Canada at a press conference on a Team Canada trade mission. Canada's Kyoto commitments "provoked concern in the West that Kyoto could be 'another NEP', that is, another unilaterally imposed policy by eastern governments over western oil and gas" (Doern and Gattinger 2002, 80). These dynamics also help to explain why the Chrétien government established the principle that no region be unfairly burdened by Kyoto implementation – a commitment needed to assuage Alberta (Doern and Gattinger 2002, 88) and one that greatly reduced Ottawa's capacity to make meaningful progress on national climate change policy. More recently, the late former Alberta premier Peter Lougheed warned in a 2007 speech to the Canadian Bar Association that federal efforts on climate change harboured significant potential for constitutional conflict if seen to impinge on Alberta's oil sands development. And in 2008, former Liberal leader Stéphane Dion's "green shift" (carbon tax) proposals in the federal election campaign vividly illustrated what happens when politicians get too close to the third rail of energy. His opponent, Prime Minister Stephen Harper, immediately compared the proposal to the National Energy Program, dismissing it as "insane" (Clark 2008).[15]

But it's not only in the environmental domain where third rail dynamics can be found. Since the signing of the two Atlantic Accords in the mid-1980s, periodic renegotiations have taken place between Ottawa and the provinces of Nova Scotia and Newfoundland and Labrador over offset payments and the treatment of provincial revenues from offshore energy development in the equalization program. Growth in provincial revenues from the offshore industry resulted in reductions in equalization payments to both provinces; by the early 2000s, roughly 70 cents of every dollar was clawed back (Dunn 2005). In 2004, during negotiations with the federal government of then Prime Minister Paul Martin over the treatment of offshore revenues in equalization, then Newfoundland and Labrador premier Danny Williams famously ordered all Canadian flags be taken down from provincial government buildings in the province, following a proposal by Martin that Williams deemed inadequate. Negotiations were eventually concluded in 2005, with the federal government bending to Williams's demands. These dramatics were to be repeated in the 2008 federal election when Premier Williams spearheaded the "Anything But Conservative" (ABC) campaign to encourage voters to withhold support for the federal Conservatives of Prime Minister Stephen Harper, whom Williams charged with reneging on a 2006 campaign promise to remove nonrenewable natural resource revenues from equalization.

In sum, the period since the mid-1980s can be termed *third rail energy federalism*: national approaches to energy or federal policies seen to impinge on provincial capacities to develop their resources are only for the brave of heart, fully prepared to suffer the consequences. So why the calls in recent years for a national energy

---

[15] Politics aside, such conflicts have given rise to much debate and analysis over the legal scope of federal jurisdiction in the climate change sphere (see, for example, Chalifour 2010; Elgie 2010; and Green 2010).

strategy and how likely are they to meet with success? The next section turns to this question.

A NATIONAL ENERGY STRATEGY FOR CANADA: GOLDEN AGE OR GOLDEN CAGE OF ENERGY FEDERALISM?

Over the last few years, numerous public, private, and non-governmental organizations at both federal and provincial levels have advocated for the development of a national energy strategy. This is a remarkable change in the Canadian energy federalism scene, which has systematically shunned national approaches to energy since the 1980s. What accounts for this turn of events? What are the profiles of these various proposals? What are their promise and prospects? This section explores these topics, arguing that energy market dynamics (changing market conditions) – the first letter of the energy MESS – are the primary driver fuelling the turn to national energy strategies. Of note, these proposals tend to be aspirational or principles-based, rather than plans for comprehensive policy harmonization in the energy sphere. And while they hold great promise to strengthen governments' capacities to address their collective and respective energy policy MESS, the prospects that a national energy strategy (or strategies) will be realized are not yet clear. Indeed, early signs suggest that Canada is unlikely to enter a golden age of energy federalism in which governments collaborate extensively and third rail energy federalism becomes a thing of the past.

An appreciation of the transformation in North American energy markets is essential to understanding why national approaches to energy have gained currency. Beginning in the late 2000s, the North American energy scene has been characterized by fundamental and rapid change. The discovery of massive reserves of unconventional oil and gas (shale oil and shale gas), combined with technological developments permitting their economic recovery (hydraulic fracturing and horizontal drilling), have transformed the energy picture in North America. The United States has proved oil reserves of 25 billion barrels, but a total of 220 billion barrels of technically recoverable resources (United States Energy Information Administration 2013).[16] In natural gas, the potential is even more striking: the US has 305 trillion cubic feet of proved reserves and over two quadrillion (2,000 trillion) in technically recoverable resources. In Canada, the change in picture is not as dramatic, particularly for oil, given the country's long-standing proved and probable/possible/speculative reserves. These are mainly in the oil sands, which have over 170 billion barrels of proved reserves, with a total of 320 technically

---

[16] Technically recoverable resources include probable, possible, and speculative reserves, so the volume must be interpreted with caution.

recoverable (United States Energy Information Administration 2013). The figures for natural gas, owing in large part to the potential of unconventional gas, are more substantial: the country has 70 trillion cubic feet of proved gas reserves (see Table 1), with another quadrillion to two quadrillion technically recoverable (United States Energy Information Administration 2013).

The so-called shale revolution in the United States has significantly changed the Canada-US energy picture: the United States has gone from being a net importer of natural gas (virtually all from Canada) to projections that it will become a net gas exporter. A similar change is under way for oil: even though the United States will continue to import oil, it may be able to significantly reduce its oil imports. The country halved oil imports between 2006 and 2012, and a February 2013 Citigroup report projected it could eliminate imports from Middle Eastern and hostile suppliers within five years[17] (Soloman 2013). Indeed, the International Energy Agency (2012) predicts that the United States will become the largest oil producer in the world by around 2020 and that its dependence on oil imports will decline from 50 percent of consumption to less than 30 percent by 2035 (76). The United States is also projected to become a net exporter of natural gas by 2035 (76). The United States Energy Information Administration (2012) has put forward similar projections, including a projected decline in net imports of natural gas from Canada between now and 2040. While long-range forecasting of energy production is notoriously challenging and US production increases of this scale may not materialize in the decades to come (particularly because some shale deposits are proving less productive than their conventional counterparts but also because conventional oil production in the US is on the decline), there is no question that the last few years have begun to call into question the size and viability of the United States as an ongoing export market for Canadian energy.

In oil, the challenge is especially daunting: given increased shale oil production and constrained refinery capacity in the US, the price of a barrel of oil in the US (West Texas Intermediate) has been selling at a discount to its European counterpart (Brent). Historically, the price spread between the two markers has been a few dollars, but it increased significantly beginning in late 2010 to between $10 and $20, or even higher (YCharts 2013). Canadian crude oil has faced a "double discount": that between West Texas Intermediate (WTI) and Brent, but also the price spread between Western Canadian Select (the price marker for oil in Western Canada) and WTI. This spread more than doubled from an average of just under ten dollars in 2009 to $21 in 2012 (Baytex 2013), hitting a whopping $42.50 in December 2012 (Els 2013). The WTI/WCS discount owes to increased US oil production and lack of sufficient pipeline capacity from Western Canada to refineries on the Gulf Coast

---

[17] Some of these changes are due to factors beyond increased US energy production, notably reduced fuel usage brought on by the economic downturn and the potential reclassification of Venezuelan oil due to the change in government.

– it is the rationale for TransCanada's Keystone XL pipeline. While price spreads have varied since 2010, ratcheting higher or lower depending upon production, consumption and refinery capacity, the volatility is especially challenging for the oil sands. Given how heavy oil from the oil sands is, it can be far more expensive to produce and refine than light/medium oil from shale formations like the Bakken basin in North Dakota, Montana, and Saskatchewan. Bakken oil has a much lower break-even point, and given its geographic location, can be an obstacle to crude from the oil sands accessing pipelines into the US market (Els 2012).

In sum, oil from the Canadian oil sands is often selling at a discount to both US and world prices and is increasingly landlocked in a hydrocarbon-rich North America. Even if the Obama administration approves the Keystone XL pipeline – which has turned out not to be, as Prime Minister Stephen Harper famously quipped, a "no-brainer" – it is not clear that this would entirely address the market challenges facing oil sands in North America or capitalize on market opportunities elsewhere. Oil sands crude would still face the discount between WTI and Brent to the extent that it persists, and if the WTI declines further, the commercial viability of oil sands projects may weaken. Indeed, Suncor and Total cancelled an upgrader planned for the oil sands in March 2013, and the dramatic drop in oil prices since mid-2014 has unleashed a flurry of major announcements from the Canadian oil patch, including layoffs, lower earnings and in some cases losses, and deferred or scaled back investments. The oil sands also face some stiff political opposition in the United States, and potential regulatory challenges in the form of low carbon fuel standards in jurisdictions like California.

So where will oil sands oil go if the opportunities to go south begin to weaken? The main alternative market opportunities are east to eastern Canadian markets, refineries, and export markets, and west to British Columbia tidewater for export to Asian markets. Both of these options are being pursued at time of writing: west via Enbridge's Northern Gateway pipeline proposal and Kinder Morgan's proposal to expand an existing pipeline into British Columbia, and east via two proposals. The first is TransCanada's Energy East pipeline, which would construct new pipelines and convert and expand an existing natural gas pipeline to carry crude oil from Alberta to refineries in Quebec and New Brunswick for Canadian and international markets. The second is the reversal of Enbridge's Line 9 pipeline between Sarnia and Montreal, which would shift from carrying imported crude from international markets to transporting western Canadian crude to Ontario and Quebec. Overall, these market dynamics are shifting the historic vertical north-south flows of energy from Canada to the United States and generating greater industrial and political interest in horizontal east-west flows within and beyond the country (Alberta is also exploring options to move oil north to Tuktoyaktuk in the Northwest Territories, to Alaska, or to Churchill Manitoba to access export markets). In natural gas, the US is likewise shifting from consumer to competitor for Canadian gas producers, and Canadian shale gas production in northeastern British Columbia in the Horn River formation has yet to secure infrastructure to the west coast for export (Lawrence 2013). In electricity, the US Energy Information Administration projections also forecast a decline in imports of Canadian power. Long-time electricity analyst Jean-Thomas Bernard doubts that electricity from Quebec and Ontario will be able to compete against power in the US Northeast produced more cheaply from shale gas (cited in Dufresne 2013).

Those advocating for a national energy strategy do so against this backdrop. Ideas to develop national approaches to energy emerged as early as 2007, when provincial premiers in the Council of the Federation laid out a plan for development of what they called a "shared vision" for energy. The premiers described the economic opportunities and importance of energy to the country, but also the challenges to its responsible development in environmental and social terms. Their "shared vision for Canada's energy future" consisted of three planks: "secure, sustainable, reliable and competitively-priced supply," "a high standard of environmental and social responsibility," and "continued economic growth and prosperity" (Council of the Federation 2007, 3). They identified a "seven point action plan" to work toward this vision, including energy efficiency/conservation; research and technology development; cleaner energy sources; safe, efficient, and environmentally responsible transportation and distribution networks; streamlined regulation; labour force development; and greater participation of the provinces and territories in international energy negotiations. Since this time, there has been a growing chorus of voices advocating for national approaches to energy, using increasingly tactical language: from the provinces' somewhat lofty sounding "shared energy vision" to national energy "frameworks" and then to "strategies" in the intervening years. A flurry of reports advocating national approaches to energy were issued by legislative, industry, environmental, and labour organizations (e.g., Blue-Green Canada 2012; Canadian Council of Chief Executives 2012; Canadian International Council 2012; Energy Policy Institute of Canada 2012; Standing Senate Committee on Energy, the Environment and Natural Resources 2012; Tides Canada 2012a, 2012b).

Many of these reports were crafted following comprehensive multiyear consultation, dialogue, and consensus-seeking processes. The Winnipeg Consensus Group (2010), a multisectoral and multistage initiative, is one of the more interesting, as it brought together representatives from a wide range of the think tank, industry, and environmental communities to discuss national approaches to energy. Spearheaded by the Business Council of Manitoba, the Canada West Foundation, and the International Institute for Sustainable Development, the process began in 2008 with an unprecedented meeting of major Canadian think tanks on the topic of energy.[18] The group included a range of research institutes (e.g., the National Roundtable on the Environment and the Economy), business groups (e.g., the Canadian Council of Chief Executives), and environmental organizations (e.g., the

---

[18] Jim Carr, president of the Business Council of Manitoba, presented on this topic at the State of the Federation 2012 Conference (November 30–December 1, 2012, in Kingston, Ontario).

Pembina Institute), which met for a series of "dialogues" to craft consensus around key planks of a "Canadian clean energy strategy" for submission to policy leaders. Tides Canada also undertook a consultation process, as did the Standing Senate Committee on Energy, the Environment and Natural Resources. And the Energy Policy Institute of Canada, comprised of major energy producing and consuming corporations, was formed for the express purpose of advocating for a national energy framework and strategy.

The bevy of reports – many of which were published over the summer of 2012, just in time for the Council of the Federation meeting in Halifax – are remarkably similar in the broad strokes of their recommendations. The majority of recommendations focus on the first imperative of the energy MESS: markets. These include fostering technology, innovation, and research (recommended in reports prepared by the Canadian Council of Chief Executives, the Canadian International Council, the Council of the Federation, the Energy Policy Institute of Canada, and Tides Canada); the need for energy infrastructure (recommended by the Canadian Council of Chief Executives, the Council of the Federation, and the Standing Senate Committee on Energy, the Environment and Natural Resources); regulatory streamlining and reform (recommended by the Canadian Council of Chief Executives, the Council of the Federation, the Energy Policy Institute of Canada, and the Standing Senate Committee); market diversification (Canadian Council of Chief Executives, Energy Policy Institute of Canada); and labour force development and jobs (Canadian Council of Chief Executives, Canadian International Council, Council of the Federation, Standing Senate Committee, and Tides Canada). Many proposals also focus on the second imperative of the energy MESS, environment, with the topics of increasing energy efficiency and conservation (Council of the Federation, Energy Policy Institute of Canada, Standing Senate Committee, Tides Canada); supporting renewable, green, or cleaner energy (Council of the Federation, Standing Senate Committee, Tides Canada); and addressing climate change (Canadian Council of Chief Executives, Canadian International Council, Energy Policy Institute of Canada, Tides Canada). In keeping with Canada's long-standing status as an energy exporter, there has been limited attention to energy security in the form of security of supply, although reliability, affordability, and critical infrastructure/cyber-security have been addressed in a number of documents. Social acceptance is mentioned in virtually all plans, with some calling for greater attention to energy literacy and the involvement of Aboriginal peoples and individual Canadians in energy development (Canadian Council of Chief Executives, Canadian International Council, Energy Policy Institute of Canada, and Tides Canada). Of note, no reports call for comprehensive energy policy harmonization: recommendations are centred on areas of policy that could benefit from federal-provincial and interprovincial collaboration.[19]

---

[19] A few reports, mainly from the business community, call for a federal policy on climate change – either a national climate change plan or a price on carbon (see, for example,

Given the extensive and broad-based interest in developing a national approach to energy and the similarity in the main thrust of recommendations coming forward from economic, social, environmental, governmental, and opinion leaders, one might expect the idea of a national energy strategy would have legs. In principle, national collaborative approaches hold great promise for Canadian governments to address their respective and collective energy MESS (e.g., collaboration on climate change adaptation and mitigation; shared approaches to labour force sourcing, training, and development; and pan-Canadian efforts on energy literacy and social acceptance). Collaborative approaches also hold the potential to strengthen a key industry of the Canadian economy, not only in its own right as a direct contributor to jobs, GDP, and trade but also as a platform for competiveness, efficiency, and growth in other industrial sectors, and a key contributor to Canadians' standard of living.

But is a return to collaboration, along the lines of the expansionist collaborative energy federalism of the 1930s to 1960s, a possibility? At time of writing, it is far from clear. The summer 2012 Council of the Federation meeting and the intervening months have underscored the challenges of moving beyond third rail energy federalism. At this meeting, discussions between the premiers on moving a national energy strategy forward got bogged down in competitive dynamics between Alberta and British Columbia over the Northern Gateway pipeline, with BC premier Christy Clark refusing to participate in the working group mandated to further the provinces' national energy strategy discussions (Council of the Federation 2012). Premier Clark used the Council meetings as a platform to underscore British Columbia's five conditions for support of the Northern Gateway pipeline: concluding the environmental review, using top-notch marine oil-spill response, deploying world-class prevention techniques for oil spills on land, sufficiently consulting and involving Aboriginal peoples, and equitably sharing revenues based on risks (CBC News 2012a, 2012b).[20] Ontario premier Kathleen Wynne and Quebec premier Philippe Couillard likewise laid out "criteria" for their support of the Energy East pipeline proposal, including that the company assess whether the project will lead to increased GHG emissions, and that TransCanada consult First Nations and local communities, take responsibility for environmental and economic risks, and take into consideration the pipeline conversion's impacts on natural gas customers (Morrow 2014). The federal government, for its part, has been all but silent on the topic of a national energy strategy. While the Harper government identified natural resource development as a key plank in its 2012 Budget (Canada 2012) and has been active on some dimensions of energy – most notably, evaluating the Chinese National Offshore Oil Company (CNOOC) takeover of Alberta-based Nexen corporation, slowly moving forward on GHG regulations for large emitters,

---

Canadian Council of Chief Executives 2012 and Winnipeg Consensus Group 2010).

[20] In November 2013 British Columbia joined the Council of the Federation's energy strategy process, as did Quebec in August 2014.

streamlining regulatory and environmental review processes, and advocating in the United States for approval of the Keystone XL pipeline – it has not engaged with the provinces as a group on energy matters. Rather, its approach to the provinces has tended to be piecemeal and based on bilateral deals, for example, federal investments in carbon capture and storage research in Alberta and loan guarantees for Newfoundland and Labrador's development of the Muskrat Falls hydroelectric project. Ottawa has also further circumscribed its role in energy development by reducing the scope of federal environmental assessments, devolving resource responsibilities to the Northwest Territories, and commencing discussions with Nunavut for energy devolution.

Although as discussed below, some progress has been made by the Council of the Federation energy working group since summer 2012, it would seem Canada is far from entering a golden age of energy federalism. But is the country doomed to remain in the "golden cage" of third rail energy federalism or can governments develop the norm of collaboration necessary for national approaches to energy?

LOOKING FORWARD: ARE WE PRISONERS OF THE (RECENT) PAST?

While the current political context does not bode well for development of a comprehensive national energy strategy, the market dynamics reorienting energy economics from north-south to east-west (or west-north) are unlikely to dissipate in the short to medium terms. Forecasting future energy supply and demand based on today's energy picture is fraught with uncertainty, but the changing energy picture in North America will continue to influence Canada-US energy relations – and in turn domestic intergovernmental energy relations – for the foreseeable future. In addition to the *M* dimension of the energy MESS, the other three energy imperatives (environment, security, and social acceptance) will likely generate greater and sustained interest in national approaches to energy, be it for putting a price on carbon that applies throughout the country, effectively addressing current and emerging threats to the physical and cyber-security of energy infrastructure, or strengthening government capacities to pursue social acceptance and support for energy resource development. These contemporary, complex, and multifaceted challenges, combined with the context of shared and sometimes overlapping jurisdiction in the energy sphere, are precisely the characteristics that call for collaboration. As Simeon (2010) notes of collaborative federalism, "the pervasive interdependence of governments faced with common policy problems means that neither level, on its own, can fully address them" (409). But will governments be able to move beyond third rail energy federalism to do so?

The longer-term prospects for a national energy strategy will ultimately depend on the willingness of federal and provincial governments to shift from assertive/

passive dynamics to a more collaborative approach. This is a major change in orientation to be sure, and the challenges of moving in this direction should not be underestimated. The "golden cage" of third rail energy federalism enables provinces to develop their resources in diverse ways reflective of local circumstances, but it militates strongly against national policy approaches – whatever their inherent merits. Prevailing norms of provincial primacy and assertiveness can blind policy-makers and citizens alike to shared energy interests across the country, serving instead to reinforce and accentuate differences and conflict. As such, instead of shooting immediately for a comprehensive "energy deal" between the federal and provincial governments, those advocating for a national energy strategy might do well to begin with a *process* and a *framework* that supports building and strengthening the norm of collaboration. This would lay the foundation for negotiation of an energy deal moving forward.

This two-step approach could begin with a framework agreement identifying the rationales, principles, and opportunities for collaboration. This would have the advantage of beginning where it matters most: developing the norm of collaboration. A framework approach could be underscored in nomenclature, by selecting a name like the Energy Collaboration Framework Agreement. The federal government should support this process as convener – not dominator – and could aim to reduce incentives for interprovincial competition and zero-sum thinking by, at a minimum, working with the provinces to establish a set of principles guiding use of the federal spending power in the energy field. These agreed-on principles would go a long way toward reducing interprovincial jealousy and bitterness over federal spending decisions that can be perceived by provinces and individual Canadians as at best discretionary, ad hoc decisions "picking energy winners" or at worst as crassly politically motivated. Without principles guiding use of the federal spending power, provinces and individual Canadians are left wondering why some provinces or projects receive support while others don't – or at least don't to the same degree. A recent example is the federal loan guarantee for Newfoundland and Labrador's Muskrat Falls development, which left both Quebec and Manitoba (if not other provinces) scratching their heads as to "why this project and not others?" Principles could include such considerations as the project's contribution to economic development, Canadians' quality of life, energy reliability, and environmental sustainability.

When it comes to process, in an ideal world, the federal government would strike a Royal Commission on Canada's Energy Future to spearhead these activities, undertake research, and ultimately make recommendations on an Energy Collaboration Framework Agreement. But the current federal government, which has clearly indicated it sees no need for a national energy strategy (Paris 2012), is unlikely to embrace such a concept. Nonetheless, intergovernmental discussions need to incorporate more fulsome engagement strategies. Given the growing importance of social acceptance, those developing an Energy Collaboration Framework Agreement must attend to its democratic underpinnings – an energy MESS in the sense of "mess hall," not disorder and disarray. Governments need to continue to move forward with executive processes but must also begin to meaningfully incorporate non-governmental actors in developing a national approach to energy. The numerous reports noted above are testament to the interest of the think tank, business, labour, and environmental communities in energy, and represent a strong foundation on which Canadian governments can build. Some progress has been made on this front by the Council of the Federation's Canadian Energy Strategy Working Group. For example, it convened a June 2013 workshop in Edmonton engaging the think tank, environmental, academic, NGO, and industry communities. In addition, when the Council of the Federation met in Charlottetown in August 2014, it tabled a revised Canadian Energy Strategy document, which explicitly incorporated the principle of "collaboration and transparency" (Council of the Federation 2014). This is a positive development to be sure, as premiers underscored the importance of collaboration between governments and with other key stakeholders including, notably, Aboriginal peoples.

Beginning with a framework approach would have the added advantage of facilitating "variable geometry" – greater levels of collaboration in some areas (and perhaps even between subsets of the full set of Canadian governments), and lesser in others. The relative weight provinces place on the four components of the energy policy MESS will vary over time, and a framework approach would accommodate differences. The process could begin by collaborating where wins can more readily be had (e.g., labour shortages and labour force development, regulatory coordination, research and technology collaboration) in order to build momentum, experience, and appetite to take on more challenging files. Where collaboration proves too challenging in the short term, governments can at a minimum share their experiences, using the "laboratory" of federalism to identify best practices on shared policy challenges (e.g., shale gas development, public engagement mechanisms, best practices for Aboriginal involvement in energy developments).

The second step would involve negotiation of an energy deal between the federal and provincial governments that would support the market access objectives of energy-producing provinces, while addressing the environmental, social, and economic concerns of other provinces, Aboriginal communities, environmental groups, and local communities. While the contours of such a deal would need to be identified through dialogue and exchange, potential elements could include mechanisms to address climate change concerns related to development of Canadian hydrocarbons (e.g., establishing some form of price on carbon applied across the country) and concerns about environmental and social risks associated with oil and gas transportation (e.g., creating a financial mechanism to address concerns that environmental risk and financial reward are not equitably distributed between producing provinces and consuming destinations when it comes to pipeline projects).

What will be imperative, though, in both of these steps, is ensuring that all four dimensions of the energy MESS are addressed and that the process reflects a mess hall – not messy – approach. Of greatest importance is the meaningful involvement of Canadians – including Aboriginal Canadians – in the process.

In sum, while it is unlikely that Canada will ever enter a golden age of energy federalism with consistent, comprehensive, and unwavering pan-Canadian collaboration, the imperatives and opportunities of the energy policy MESS are such that governments need to break with (recent) tradition and move beyond third rail energy federalism. National approaches to energy have been on the table in the past and need to become politically acceptable going forward.

REFERENCES

Anderson, George. ed. 2012. *Oil and Gas in Federal Systems.* Toronto: Oxford University Press.

Baytex. 2013. "WCS Pricing." January 2013. http://www.baytex.ab.ca/files/pdf/Operations/Historical%20WCS%20Pricing_January%202013.pdf.

Blue-Green Canada. 2012. *More Bang for Our Buck: How Canada Can Create More Energy Jobs and Less Pollution.* Toronto: Blue-Green Canada.

Canada. 2012. *Jobs Growth and Long-Term Prosperity, Economic Action Plan 2012.* Ottawa: Her Majesty the Queen in Right of Canada.

Canadian Association of Petroleum Producers. 2013. *Technical Report: Statistical Handbook for Canada's Upstream Petroleum Industry.* June. Calgary: CAPP.

Canadian Council of Chief Executives. 2012. *Framing an Energy Strategy for Canada: Submission to the Council of the Federation.* Ottawa: Canadian Council of Chief Executives.

Canadian Electricity Association. 2013. *Key Canadian Electricity Statistics.* Released May 21. http://www.electricity.ca/media/IndustryData/KeyCanadian ElectricityStatistics21May2013.pdf.

Canadian International Council. 2012. *The 9 Habits of Highly Effective Resource Economies.* Toronto: Canadian International Council.

CBC News. 2012a. "B.C. Seeks 'Fair Share' in New Gateway Pipeline Deal: Province Lays Out 5 Criteria for Provincial Approval of All New Crude Pipelines." Posted July 23. http://www.cbc.ca/news/politics/story/2012/07/23/pol-bc-pipeline-clark-gateway.html.

—. 2012b. "B.C. Premier Boycotts National Energy Strategy: Christy Clark Refuses Participation without Deal on Northern Gateway Pipeline." Posted July 27. http://www.cbc.ca/news/politics/story/2012/07/27/pol-premiers-friday.html.

Chalifour, Nathalie. 2010. "The Constitutional Authority to Levy Carbon Taxes." In *Canada: The State of the Federation 2009. Carbon Pricing and Environmental Federalism,* edited by Thomas J. Courchene and John R. Allan, 177-95. Montreal and Kingston: McGill-Queen's University Press.

Clark, Campbell. 2008. "Dion's Carbon Tax Would 'Screw Everybody,' PM Says." *Globe and Mail,* June 21. http://www.theglobeandmail.com/news/national/dions-carbon-tax-would-screw-everybody-pm-says/article675145/.

Council of the Federation. 2007. *A Shared Vision for Energy in Canada*. August. Ottawa: Council of the Federation.

—. 2012. "Premiers Guide Development of Canada's Energy Resources." Press release, July 27, Halifax.

—. 2014. "Canadian Energy Strategy." Press release, August 29, Charlottetown.

Courchene, Thomas J. 2004. "Confiscatory Equalization: The Intriguing Case of Saskatchewan's Vanishing Energy Revenues." *Choices* 10 (2). Montreal: Institute for Research on Public Policy.

Courchene Thomas J., and John R. Allan, eds. 2010. *Canada: The State of the Federation 2009. Carbon Pricing and Environmental Federalism*. Montreal and Kingston: McGill-Queen's University Press.

Doern, G. Bruce, and Monica Gattinger. 2002. "Another 'NEP'? The Bush Energy Plan and Canada's Political and Policy Responses." In *Canada among Nations 2002: A Fading Power*, edited by Norman Hillmer and Maureen Appel Molot, 74-96. Toronto: Oxford University Press.

—. 2003. *Power Switch: Energy Regulatory Governance in the 21ˢᵗ Century*. Toronto: University of Toronto Press.

Doern, G. Bruce, and Glen Toner. 1985. *The Politics of Energy: The Development and Implementation of the NEP*. Toronto: Methuen.

Dufresne, Jean-Marc. 2013. "Against the Current? The Quebec Government Is Relying on the Redevelopment of Hydroelectric Power to Boost Its Revenues. Yet the Economic Outlook for This Source of Energy Isn't Rosy, says Jean-Thomas Bernard." *Research Perspectives: A Journal of Discovery and Innovation from the University of Ottawa* (Summer): 14-15.

Dunn, Christopher. 2005. "Why Williams Walked, Why Martin Balked: The Atlantic Accord Dispute in Perspective." *Policy Options* 26 (2): 9-14.

Elgie, Stewart. 2010. "Carbon Emissions Trading and the Constitution." In *Canada: The State of the Federation 2009. Carbon Pricing and Environmental Federalism*, edited by Thomas J. Courchene and John R. Allan, 161-76. Montreal and Kingston: McGill-Queen's University Press.

Els, Frik. 2012. "Fire Sale: Oil Sands Players Now Get $45 a Barrel vs Global Price of $109." *Mining.com*, December 14. http://www.mining.com/oil-sands -players-now-get-only-45-a-barrel-vs-global-price-of-109-54211/.

—. 2013. "Welcome to Canada, Land of the $63 Barrel of Oil." *Mining.com*, March 14. http://www.mining.com/welcome-to-canada-land-of-the-63-barrel-of-oil-90773/.

Energy Policy Institute of Canada. 2012. *A Canadian Energy Strategy and Framework: A Guide to Building Canada's Future as a Global Energy Leader*. Ottawa: Energy Policy Institute of Canada.

Environment Canada. 2014. *Greenhouse Gas Emissions Data*. Last modified April 11. https:// www.ec.gc.ca/indicateurs-indicators/default.asp?lang=cn&n=BFB1B398-1.

Froschauer, Karl. 1999. *White Gold: Hydroelectric Power in Canada*. Vancouver: UBC Press.

Gattinger, Monica. 2005. "Canada-United States Electricity Relations: Policy Coordination and Multi-Level Associative Governance." In *How Ottawa Spends 2005–2006: Managing the Minority*, edited by G. Bruce Doern, 143-62. Montreal and Kingston: McGill-Queen's University Press.

—. 2012. "Canada-United States Energy Relations: Making a MESS of Energy Policy." *American Review of Canadian Studies* 42 (4): 460-73.

Gattinger, Monica, and Geoffrey Hale, eds. 2010. *Borders and Bridges: Canada's Policy Relations in North America*. Toronto: Oxford University Press.

Green, Andrew. 2010. "Carbon Pricing, the WTO and the Canadian Constitution." In *Canada: The State of the Federation 2009. Carbon Pricing and Environmental Federalism*, edited by Thomas J. Courchene and John R. Allan, 197-217. Montreal and Kingston: McGill-Queen's University Press.

International Energy Agency. 2012. *World Energy Outlook 2012*. Paris: OECD/IEA.

Jaccard, Mark, and Nic Rivers. 2007. "Canadian Policies for Deep Greenhouse Gas Reductions." In *A Canadian Priorities Agenda: Policy Choices for Economic and Social Well-Being*, edited by Jeremy Leonard, Christopher Ragan, and France St-Hilaire, 75-106. Montreal: Institute for Research on Public Policy.

Janigan, Mary. 2012. *Let the Eastern Bastards Freeze in the Dark: The West Versus the Rest since Confederation*. Toronto: Knopf Canada.

Johnston, Peter F. 2008. *Oil and Terrorism: Al Qaeda's Threat*. Defence R&D: Centre for Operational Research and Analysis. DRDC CORA TM 2008-012 April. Ottawa: Minister of National Defence, Her Majesty the Queen in Right of Canada.

Lawrence, Daina. 2013. "Canada Faces Hurdles Joining Shale Gas Revolution." *Globe and Mail*, 21 February. http://www.theglobeandmail.com/report-on-business/industry-news/energy-and-resources/canada-faces-hurdles-in-joining-shale-gas-revolution/article8900288/.

McDougall, John. 1982. *Fuels and the National Policy*. Toronto: Butterworth.

Milne, David. 1986. *Tug of War: Ottawa and the Provinces under Trudeau and Mulroney*. Toronto: James Lorimer.

Morrow, Adrian. 2014. "Premiers Wynne and Couillard Set Seven Criteria for Energy East." *Globe and Mail,* November 21. http://www.theglobeandmail.com/news/politics/premiers-wynne-and-couillard-set-seven-criteria-for-energy-east/article21714915/.

Panitch, Leo. 1977. *The Canadian State: Political Economy and Political Power*. Toronto: University of Toronto Press.

Paris, Max. 2012. "No Need for National Energy Strategy Oliver Says." *CBC News: Politics*, September 11. http://www.cbc.ca/news/politics/story/2012/09/11/pol-oliver-national-energy-strategy.html.

Plourde, André. 2005. "The Changing Nature of National and Continental Energy Markets." In *Canadian Energy Policy and the Struggle for Sustainable Development*, edited by G. Bruce Doern, 51-82. Toronto: University of Toronto Press.

—. 2012. "Canada." In *Oil and Gas in Federal Systems*, edited by George Anderson, 88-120. Toronto: Oxford University Press.

Pratt, Larry. 1976. *The Tar Sands: Syncrude and the Politics of Oil*. Edmonton: Hurtig Publishers.

Richards, John, and Larry Pratt. 1979. *Prairie Capitalism: Power and Influence in the New West*. Toronto: McClelland and Stewart.

Sanger, David, David Barboza, and Nicole Perlroth. 2013. "Chinese Army Unit Is Seen as Tied to Hacking against U.S." *New York Times*, February 18. http://www.nytimes.com/2013/02/19/technology/chinas-army-is-seen-as-tied-to-hacking-against-us.html?pagewanted=all&_r=0.

Simeon, Richard. 2010. "Federalism and Intergovernmental Relations." In *The Handbook of Canadian Public Administration,* 2nd ed., edited by Christopher Dunn, 401-21. Toronto: Oxford University Press.

Soloman, Lawrence. 2013. "Fight Jihad, Stop Carbon Taxes." *Financial Post*, February 21. Last modified February 22. http://opinion.financialpost.com/2013/02/21/lawrence-solomon-shale-means-security/.

Standing Senate Committee on Energy, the Environment and Natural Resources. 2012. *Now or Never: Canada Must Act Urgently to Seize Its Place in the New Energy World Order*. Ottawa: Senate of Canada.

Statistics Canada. 2013. "Gross Domestic Product, Expenditure-Based, by Province and Territory." Last modified December 20. http://www.statcan.gc.ca/tables-tableaux/sum-som/l01/cst01/econ15-eng.htm.

—. 2014. "Population by Year, by Province and Territory." Last modified September 26. http://www.statcan.gc.ca/tables-tableaux/sum-som/l01/cst01/demo02a-eng.htm.

Tides Canada. 2012a. *A New Energy Vision for Canada*. March. n.p.: Tides Canada.

—. 2012b. *Towards a Clean Energy Accord: How and Why a Canadian Energy Strategy Can Accelerate the Nation's Transition to a Low-Carbon Economy*. June. n.p.: Tides Canada.

United States Energy Information Administration. 2012. *Annual Energy Outlook 2013 Early Release Overview*. Report Number DOE/EIA-0383ER (2013). Washington, DC: United States Energy Information Administration.

—. 2013. *International Energy Statistics* (Petroleum and Natural Gas Reserves). http://www.eia.gov/cfapps/ipdbproject/IEDIndex3.cfm.

Winnipeg Consensus Group. 2010. *Towards a Canadian Clean Energy Strategy: Summary of the Banff Clean Energy Dialogue, April 8–10, 2010*. n.p.: Winnipeg Consensus Group.

YCharts. 2013. *Brent WTI Spread*. http://ycharts.com/indicators/brent_wti_spread.

5

SURPLUS RECYCLING AND THE CANADIAN FEDERATION: THE HORIZONTAL FISCAL BALANCE DIMENSION

Thomas J. Courchene

INTRODUCTION

In his 2011 book, *The Global Minotaur: America, the True Origins of the Financial Crisis and the Future of the World Economy,* Greek economist Yanis Varoufakis makes a convincing case that effective surplus-recycling mechanisms (SRMs) are essential for maintaining the internal stability and resiliency of macroeconomic systems. While his analysis is insightful and persuasive, the role of surplus-recycling mechanisms has long been centre stage in ensuring international macroeconomic equilibrium. Arguably the most familiar SRM was the "rules of the game" under the gold standard or, more instructively, under the price-specie-flow mechanism. Countries running balance-of-payments surpluses will experience inflows of gold (specie) that in turn will increase domestic wages and prices thereby eroding their balance-of-payments surplus by decreasing exports and increasing imports. Balance-of-payments-deficit countries will experience the opposite impacts, with the result that the system will re-equilibrate. However, if the balance-of-payments-surplus countries sterilize the gold or specie inflow, then the surplus-recycling

---

The statistical analysis in this chapter relates to the latest data available at the time of writing and at the 2012 State of the Federation conference. With the recent collapse of energy prices, the cross-province comparisons will be very different, especially for the energy-rich provinces. However, the challenges that are outlined in the chapter will reappear if and when energy prices recover.

mechanism is stymied and the burden of adjustment is shifted to the deficit countries in the form of austerity or exchange rate depreciation, both of which significantly increase the political and economic adjustment costs, and undermine the principles of the system.

By way of an ongoing example, Varoufakis argues that the euro version of an SRM likewise undermines the entire system. In effect, while Germany runs an overall surplus with the other euro countries (especially those in the southern core), it invests these surpluses in the dollar area (including Asia), not in the euro area. This perpetuates the euro-related German surpluses and effectively transfers the adjustment back onto the deficit euro countries, an adjustment that without access to exchange rate depreciation will almost certainly exacerbate the likelihood of recovery in the short term.

The US-China relationship presents another example, one where China's trade surpluses are indeed cycled back to the United States but in a manner that has served to perpetuate the US fiscal and balance-of-payments challenge. Specifically, China has pegged its yuan to the greenback, and in spite of its huge trade surplus with the United States it has essentially maintained the peg. However, this requires China to become the buyer of last resort of any and all US treasuries, which, in turn, effectively removes the US budget constraint and serves to entice the United States to defer setting its fiscal house in order, even to the point where US indebtedness is now endangering its very economic future. Readers will recognize that China's approach is a modern version of reneging on the gold standard "rules of the game."

With this as brief backdrop relating to the concept of surplus recycling, my ongoing research thrust is to identify and assess the efficacy of Canada's surplus-recycling mechanisms as they relate to interprovincial and federal-provincial fiscal and economic stability. This chapter directs attention to the interprovincial rather than the federal-provincial dimension and in particular to two aspects of what has come to be referred to as horizontal fiscal balance.

The first of these is the constitutionally mandated equalization program that is designed to ensure that the fiscally weak provinces have access to revenues sufficient to mount reasonably comparable provincial public goods and services. The key conclusion here will be that the equalization is too generous to the traditional have-not provinces (Prince Edward Island, Nova Scotia, New Brunswick, Quebec, and Manitoba) whereas it falls well short with respect to Ontario.

The second surplus-recycling system, or lack thereof, focuses on the high-fiscal-capacity provinces, essentially the resource-rich provinces. Here the challenge is to ensure that these provinces do not veer too far off in the direction of becoming tax havens and/or providers of superior provincial public goods and services. However, simmering just below the surface are several complex and loaded policy issues – ensuring that a hydrocarbon/hydroelectric economy will benefit all of Canada, managing the so-called Dutch disease and the associated manufacturing resources tug-of-war, and stewarding resources to ensure that they will benefit future as well as current generations, among others.

Readers will recognize that these are highly explosive issues: they tamper, albeit indirectly, with provincial entitlements; they are inherently zero-sum games; they embody empirical assessments that are both complex and controversial, and so on. Phrased differently, there can be no first-best solutions. As such, the policy recommendations cannot consist of doctrinaire remedies, but instead must of necessity take the form of a series of options or avenues for improving the operations of these macro-equilibrating mechanisms. Indeed, the primary contribution of the chapter may well lie not in providing solutions, but rather in shedding political and empirical light on some existing inadequacies of the status quo in respect of the ability of these SRMs to provide the resilience and stability that the Canadian federation requires.

EQUALIZATION AS AN INTERPROVINCIAL SURPLUS-RECYCLING MECHANISM

There are many programs that recycle revenues/incomes/benefits across individuals and provinces. Employment Insurance serves to transfer benefits at the individual level from the employed to the unemployed, and at the interprovincial level (via the special regional provisions relating to entry qualifications and benefit duration) from low-unemployment provinces to high-unemployment provinces. Progressive tax systems combined with proportional expenditure systems transfer benefits from high-income individuals (and provinces) to lower-income individuals (and provinces). Within each province, tax-financed health care transfers benefits from rich and healthy citizens to poor and unhealthy citizens.

However, important as these transfers are in terms of creating an equitable and resilient society, they are interpersonal or interprovincial transfers to *persons*, whereas the surplus-recycling mechanisms that are the focus of this chapter relate to transfers and other policy measures designed to address horizontal fiscal imbalances across *provinces*. While transfers like the CHT (Canada Health Transfer) and the CST (Canada Social Transfer) would qualify as SRMs – indeed one of the policy options flowing from the ensuing analysis is that these transfers might play an even larger surplus-recycling role in the future – the dominant interprovincial surplus-recycling mechanism is the equalization program, to which the analysis now turns.

Philosophy and Principles

From Section 36(2) of the Constitution Act, 1982:

> Parliament and the government of Canada are committed to the principle of making equalization payments to ensure that provincial governments have sufficient revenues to provide reasonably comparable levels of public services at reasonably comparable levels of taxation.

Canada's system of equalization payments was introduced as part of the 1957 Tax Sharing Arrangements that transferred shares of federal taxes back to the provinces – 10 percent of federal income taxes, 9 percent of federal corporate income taxes, and 50 percent of succession duties. Since these federal abatements were allocated to the provinces on a *derivation basis* (i.e., on the basis of what was actually collected in the respective provinces), larger per capita revenues were generated in the richer provinces. Ottawa responded by introducing an equalization program to offset some of these per capita differences. From the outset, equalization payments have always been unconditional transfers in that the recipient provinces can spend them as they please.

While equalization payments are a key component of interprovincial surplus recycling, this recycling does not involve direct transfers of provincial revenues from rich to poor provinces. Rather, Ottawa makes these payments to the poorer provinces from its consolidated revenue fund (CRF). Although identically situated citizens, no matter where they reside, will contribute the same amount to the CRF, this nevertheless means that, in aggregate, residents of rich provinces will pay higher per capita revenues to Ottawa than will residents of poorer provinces. In this sense it can be said that Albertans (but not Alberta) contribute more per capita to the cost of equalization than say, residents of Nova Scotia, but this is also the case for National Defence or Old Age Security or any other federal spending program.

A final and often-overlooked point merits airing, namely, that equalization also benefits the richer provinces. Specifically, without the presence of an equalization program, there is no way that Canada would be as decentralized on the taxation front as we currently are, which clearly and hugely privileges the rich provinces.

Attention is now directed to the operations of the equalization program since fiscal year 2005–06, after which focus will turn to the performance of equalization in fiscal year 2012–13.

The Recent Evolution of Canada's Equalization Program

Table 1 presents per capita data on equalization (and the associated payments from the Offshore Accords) from fiscal year 2005–06 through to 2012–13. It may come as a surprise that at one time or another over this period all provinces except Alberta have been recipients of equalization.[1] While Newfoundland's revenues from offshore energy have made it a "have" (i.e., non-equalization-receiving)

---

[1] In the early years of equalization Alberta qualified for equalization, and in the mid-1980s the province also qualified for several hundred million dollars of "stabilization payments." No longer existing, these payments were designed to ensure that a province's revenues (at unchanged tax rates) would not decline from one year to the next.

Table 1: The Evolution of Equalization Payments (2005–2013, $ million)

	2005–06	*2006–07*	*2007–08*	*2008–09*	*2009–10*	*2010–11*	*2011–12*	*2012–13*
NL								
Equalization	861	687	477	0	0	0	0	0
OA	(189)	(329)	(494)	(557)	(465)	(642)	(536)	0
PEI	277	291	294	322	340	330	329	337
NS								
Equalization	1,344	1,386	1,465	1,465	1,391	1,110	1,167	1,268
OA	(31)	(57)	(68)	(106)	(180)	(227)	(250)	(458)
NB	1,348	1,451	1,477	1,584	1,689	1,581	1,483	1,495
QC	4,798	5,539	7,160	8,028	8,355	8,552	7,815	7,391
ON	0	0	0	0	347	972	2,200	3,261
MB	1,601	1,709	1,826	2,063	2,063	1,826	1,666	1,671
SK	89	13	226	0	0	0	0	0
AB	0	0	0	0	0	0	0	0
BC	590	459	0	0	0	0	0	0
YT TFF	501	517	544	564	612	653	705	767
NT TFF	737	757	843	805	864	920	996	1,070
NU TFF	821	844	893	944	1,022	1,091	1,175	1,273
CA								
Equalization	10,907	11,535	12,925	13,462	14,185	14,372	14,659	15,423
OA	(220)	(386)	(562)	(663)	(645)	(869)	(786)	(458)
TFF	2,058	2,118	2,279	2,313	2,498	2,664	2,876	3,111

Note: OA = Offshore Accord; YT = Yukon Territory; NT = Northwest Territories; NU = Nunavut; TFF = Territorial Formula Financing.

Source: Department of Finance Canada (2013).

province since 2008–09, it still received significant Offshore Accord[2] payments until 2011–12. On the other hand, Nova Scotia's offshore rebates have grown every year, with presumably larger increases to come.

However, the most policy significant economic news in the table is the descent of Ontario into the ranks of the receiving provinces. In fiscal year 2012–13, its $3.261 billion equalization payment represents 21 percent of total equalization ($15.423 billion, from the last row panel of the table). Given this rapid rise in populous Ontario's equalization entitlements in the presence of an overall cap on the system, the inevitable result was that all other recipient provinces saw declines in their annual entitlements from what they otherwise would have been (Nova Scotia is an exception if one includes its offshore payments). Not surprisingly, this has led to major concerns on the part of the traditional recipients, even to the point of pressing Ottawa to prevent Ontario's entitlements from having a negative impact on their own entitlements.[3] The more general point here is that equalization is becoming increasingly problematical when six provinces with over 70 percent of the all-provinces population fall into the have-not category: without the equalization cap, each net additional dollar accruing to a have province would lead to an increase in equalization of 70 cents!

Prior to turning to a more detailed description of the workings of the program, it should be noted that the data in Table 1 for the three territories (referred to as Territorial Formula Financing) reveal that the entitlements for all three have risen each and every year. In what follows no further attention will be devoted to Territorial Formula Financing.

The Anatomy of Provincial Finances: 2012–13

Table 2 presents per capita data relating to an overview of provincial finances for fiscal year 2012–13 based on the December 16, 2011, equalization estimates.[4] The

---

[2] The Offshore Accords for Newfoundland and Labrador and Nova Scotia were designed to refund an equalization reduction arising from offshore energy revenues accruing to these provinces. This led to the highly unfair situation in 2001–02 and thereabouts where these two provinces had zero equalization clawbacks on their energy whereas then-have-not province Saskatchewan endured confiscatory clawbacks. See Courchene (2004).

[3] While Ottawa has thus far not acted on this concern, as partial protection it did agree to introduce a TTP (Total Transfer Protection) program that ensures that a province's current level of total transfers (CHT/CST, equalization, and the prior year's TTP) are no lower than those in the prior year. The moneys associated with TTP are not factored into the tables in this chapter.

[4] The equalization payments in Table 1, when converted to a per capita basis, differ somewhat from those in Table 2. Although they are both official estimates, they were produced at different dates and with slightly different data availability.

overall total depicted in row 7 ("overall fiscal capacity") is the sum of equalization-defined fiscal capacity including all resource revenues plus the Offshore Accords plus the CHT/CST transfers.

In more detail, row 1 contains the provincial per capita values for fiscal capacity that enter the equalization formula. These are the sum of 100 percent of the per capita fiscal capacity for the components of four non-energy tax bases (i.e., personal income taxes, corporate income taxes, sales taxes, and property taxes) plus 50 percent of the actual values for resource revenues.[5] In turn, the fiscal capacity estimates for the four non-energy tax bases for each province are the product of the value of the province's tax base multiplied by the national average tax rate (not the province's own rate).

Row 2 contains the per capita values of equalization payments. If equalization payments were strictly formula driven, these values would be the difference (where positive) between the all-Canada average for fiscal capacity in the final column of row 1 of Table 2 and the respective provincial fiscal capacities. However, overall equalization payments were scaled down (on an equal per capita basis) to ensure that they would be in line with the overall cap on equalization.

Row 3 then adds the other 50 percent of resource revenues (where relevant),[6] and row 4 contains the Offshore Accord offset payment for Nova Scotia. Row 5 sums the previous four rows to obtain what might be called the aggregate taxation-cum-equalization measures of provincial fiscal capacity. Row 6 then contains the equal per capita federal-provincial CHT/CST transfers (i.e., about $1,200 per capita).[7] Finally, row 7 is the sum of row 5 and row 6 and, as such, represents the overall per capita fiscal capacity available to the individual provinces. Note that this is not quite the same as *actual* overall per capita provincial revenues because it assumes that provinces levy taxes at all-province-average tax rates, among other reasons to be elaborated later.

These overall or aggregate fiscal capacities vary from roughly $8,300 per capita for the six equalization-receiving provinces to $8,992 for British Columbia, $11,086 for Saskatchewan, $12,307 for Newfoundland and Labrador and, finally, $13,930 for Alberta. These are dramatic differences, so much so that if they persist the likely outcome will surely lead to superior public goods and/or tax havens in the high fiscal capacity provinces, about which more will be detailed later.

---

[5] Only 50 percent of resource revenues enter the current equalization program.

[6] The sum of the 50 percent of resource royalties is $11,281.5 million. Hence total resource royalties associated with the 2012–13 equalization calculations is $22,563 million.

[7] Actually, Alberta will only receive the same per capita amount of these federal-provincial transfers as the other provinces beginning in 2014–15, but this is ignored for purposes of this table. A revenue-testing approach to these transfers will be developed later in this chapter.

Table 2: The Anatomy of Provincial Finances
(2012–13, $/capita)

	NL	PEI	NS	NB	QC	ON	MB	SK	AB	BC	CAN Ave
1. Fiscal capacity	8,444	4,711	5,501	5,097	6,036	6,840	5,721	8,466	11,351	7,453	7,174
2. Equalization	0	2,377	1,347	1,993	943	249	1,368	0	0	0	
3. Other 50% of resource revenues	2,663	2	164	54	188	9	73	1,420	1,379	339	
4. Offshore Accords			155								
5. Subtotal (1–4)	11,107	7,090	7,167	7,144	7,167	7,098	7,162	9,886	12,730	7,792	7,509
6. CHT/CST	1,200	1,200	1,200	1,200	1,200	1,200	1,200	1,200	1,200	1,200	
7. Overall fiscal capacity	12,307	8,290	8,367	8,344	8,367	8,298	8,362	11,086	13,930	8,992	8,709

Note: CHT = Canada Health Transfer; CST = Canada Social Transfer.

Source: Department of Finance, Equalization estimates, December 16, 2011.

An Alberta Detour

A brief detour is in order because many readers will be puzzled by the fact that super-rich Alberta with its near-$14,000 in per capita revenues is struggling with a significant *deficit* more recently. Not surprisingly, the principal reason for this is the sharp reduction in expected energy revenues. However, there are at least three additional reasons for the difference between Alberta's superior fiscal capacity in Table 2 on the one hand and its current deficit woes on the other. The first relates to the earlier-noted "lag" that is built into the equalization formula and calculations, namely, a three-year average (with weights of 50-25-25 respectively) lagged two years. Thus, for fiscal year 2012–13 this means that the data entering the equalization formula are as follows – 50 percent of the 2010–11 data, 25 percent of the 2009–10 data, and 25 percent of the 2008–09 data. Hence, the high oil price of 2008 (including the spike to $150 per barrel) has a 25 percent weight in the Table 2 results. This demonstrates a problem with the current equalization program: in this volatile world-energy-price environment, the equalization authorities should surely take steps to ensure that the formula embodies more up-to-date data.

The second reason for the difference between Alberta's fiscal capacity and its deficit challenges is that the data that do enter the formula relate to a province's *fiscal capacity and not to its actual fiscal revenues*. If a province opts not to tax one of its revenue sources, it still will be assigned its relevant fiscal capacity. This is especially relevant for Alberta – it obviously has a fiscal capacity for generating sales tax revenues, but it chooses not to levy such a tax. Thus the Table 2 fiscal capacity data for Alberta in row 1 include what it *could raise, not what it did raise* from a provincial sales tax levied at national-average tax rates. Currently, the value of the sales tax entry (and therefore the actual value of the foregone revenue) for Alberta is reported to be in the neighbourhood of $6 billion.

The third reason has already been alluded to, namely, that Alberta has been moving in the direction of becoming both a tax haven (e.g., it has no provincial sales tax, as just noted, and it has Canada's lowest personal income tax) and a provider of superior public goods. In terms of the latter, the Fraser Institute's Mark Milke points out that Alberta has some of the highest per capita program expenditures of any of the provinces, including paying its teachers 20 percent more than in other provinces (cited in Gerson 2012).

This Alberta detour aside, the role of the next two sections is to make the case that the conceptual basis of Canada's approach to equalization – ensuring that the receiving provinces end up with approximately the same *nominal* (actual) per capita dollars – should be reconsidered. Specifically, the analysis will focus on the implications of ignoring the cost/price differentials in the provision of provincial public goods and services on the one hand and the differential fiscal needs across the various provinces on the other.

Capitalization and Equalization

Alone among mature federations, the United States has no formal revenue-equalization program. No doubt part of the reason relates to the reality that point-in-time income distribution is not a high priority in the United States. Indeed, as detailed in my *Rekindling the American Dream* (2011), the United States has the most unequal distribution of income in the rich nations' club, so it is perhaps not much of a stretch that this mentality underpins the indifference to the distribution of per capita revenues across American states.

However, Wallace Oates (1983), one of the leading experts on US federalism, rationalizes the absence of an equalization program in economic terms. Specifically, while it is certainly true that states with higher per capita incomes have the potential for having correspondingly higher per capita revenues, these differential state incomes will be capitalized in terms of higher wages and rents that, in turn, will increase the costs/prices of producing state-level public goods and services.

Similarly, lower-income states will also have lower wages and rents so that they do not need the higher level of per capita revenues of the higher income and revenue states in order to provide comparable levels of public goods and services. In other words, income differentials will be "capitalized" in terms of wages and prices, so that in the final analysis there may be little or nothing to equalize, as it were. In Oates's view, the decision to have an equalization program in a federal system is more a matter of "taste" than of social or economic principles.

Within the Oates framework and under the assumption of full or 100 percent capitalization, there would be no need for an equalization program since higher wages and rents in high-income states would offset their revenue advantage and vice versa for states with low wages and rents. However, most analysts would take the view that the assumption of 100 percent capitalization is extreme. But so is the opposite assumption that Canada embraces in its equalization program, namely, that there is *zero capitalization* so that one can ignore the prices/costs of provincial public goods and services in the calculation of equalization payments. This issue merits further attention.

An excellent place to start this rethinking is with the Constitution. Section 36(2), reproduced earlier, does not call for equalizing per capita revenues across the recipient provinces. Rather the stated thrust is that provinces should end up with revenues sufficient to provide *reasonably comparable **levels** of public goods and services*. This being the case, the ability of the recipient provinces to provide comparable levels or bundles of public goods and services will obviously depend not only on provincial revenues but, as well, on the prices or costs of providing these bundles. In other words, my reading of 36(2) is that equalization is about providing comparable *quantities,* that is, about providing comparable *real or purchasing-power-corrected* bundles of provincial public goods and services.

What difference would this approach make to the results in Table 2? Table 3 provides one answer. Row 1 of the table reproduces the overall per capita fiscal capacity figures from row 7 of Table 2. Note that these figures *include* equalization. In order to convert these figures to real or purchasing-power bundles of public goods and services, we need an index of the prices/costs by province of these goods and services. Thankfully, Peter Gusen (2012a, 2012b) has calculated just such an index for fiscal year 2008–09 in connection with his path-breaking Mowat Centre research directed toward measuring expenditure needs across provinces. His indexes by province (with Canada equal to 1.00) appear as row 2 in Table 3. The indexes are calculated as the prices/costs of a weighted average of six categories of public goods and services: wages and salaries,[8] transfers, construction contracts,

---

[8] In order that provinces not be able to influence their own equalization, the choice for the measure of wages and salaries relates not to public sector wages but rather to the index of average industrial earnings in the relevant province.

Table 3: **Capitalization and Equalization:**
Incorporating Wages and Prices into Equalization
(2012–13)

	NL	PEI	NS	NB	QC	ON	MB	SK	AB	BC	CA
1. Total provincial revenues[a] ($ per capita)	12,307	8,290	8,367	8,344	8,367	8,298	8,362	11,086	13,930	8,992	8,709
2. Gusen's price index (2009) for public goods and services[b]	.969	.895	.934	.943	.959	1.020	.964	.976	1.089	.986	1.00
3. Purchasing power bundles row 1 ÷ row 2	12,700	9,263	8,955	8,848	8,725	8,135	8,674	11,359	12,792	9,117	
4. Equalization from row 2, Table 2, $/cap	0	2,377	1,347	1,993	943	249	1,368	0	0	0	
5. Equalizing real fiscal capacities $/capita[c]	0	1,832	1,206	1,691	800	388	1,053	0	0	0	

Notes:

[a] Reproduced from row 7 of Table 2.

[b] Gusen's (2012b) weighted average of the price of public goods and services incorporates wages and salaries, transfers, construction contracts, health-care purchases, consulting services, and "others" (7–8).

[c] These equalization entitlements are calculated as follows: (1) convert the row 1, Table 2 per capita fiscal capacity values to real terms by dividing them by the Gusen indexes (row 2 of this table); and (2) bring all the recipient provinces up to the level consistent with exhausting the maximum allowable amount of equalization. Consult the relevant text for additional comments.

Source: Table 2 and author's calculations.

health-care purchases, consulting services, and a residual category referred to as
"other." Row 3 is obtained by dividing row 1 by row 2 where the resulting values
represent estimates of the real purchasing power of post-equalization aggregate
provincial revenues.[9]

The results border on the astounding. Ontario, with $8,135 per capita in real
purchasing-power revenues, comes off as the most fiscal-capacity-deprived province, and by a considerable margin.[10] The next closest are Manitoba with $8,674
and Quebec with $8,725. Lest one think that these are small differences, with a
population in the neighbourhood of 13 million Ontario's near-$600 per capita
shortfall (in real terms) relative to Quebec means that it would take roughly $8
billion (of real purchasing power) to close the Ontario-Quebec gap. Moreover, and
equally revealing, non-equalization-receiving province British Columbia ends up
with a lesser ability to provide per capita real quantities of public goods than does
equalization-receiving Prince Edward Island, traditionally viewed as the province
with the smallest fiscal capacity.

At one level, row 3 of Table 3 is the appropriate vantage point for assessing
the adequacy of Canada's equalization system in terms of the overall distribution
of per capita revenues across the recipient provinces. From this perspective, the
equalization system is clearly failing Ontario as an effective surplus-recycling
mechanism. Although Ontario's descent into have-not status is in part a relative
decline (resource-rich provinces have faired much better) as well as an absolute
decline (Ontario has been hurt by US offshoring of manufacturing and by the
Dutch disease, on which more later), the province cannot escape the reality that
some sizeable share of its ongoing fiscal woes can be traced to inappropriate policy
decisions in several key areas, for example, electricity.

Since the comparison in row 3 of Table 3 includes the equalization payments from
row 2 of Table 2, a more appropriate comparison might be to calculate equalization
on the basis of purchasing-power-adjusted provincial fiscal capacities. This would
involve dividing provincial fiscal capacities in row 1 of Table 2 by the Gusen price/
cost index in row 2 of Table 3. Then one would transfer equalization dollars to the
lowest purchasing-power-adjusted fiscal-capacity province until it is brought up to
the second lowest, and then transfer equalization dollars to these two provinces until
they achieve the level of the third lowest, and so on until the allowable equalization

---

[9] Note that Gusen does not undertake this sort of exercise. Rather, he uses these price/
cost data as part of the calculation of what he calls "expenditure needs." Indeed, he does
not even present data on the overall index that appears in row 2 of Table 3, although he
does provide the data necessary for one to calculate the index, which I have done. More on
Gusen's expenditure-needs approach in the next section.

[10] Ontario also comes off with the lowest real fiscal capacity if one incorporates capitalization via an index of average weekly wages for public administration and health/social
services (Courchene 2008, Table 1).

pool runs dry. This process leads to the per capita equalization payments in row 5 of Table 3, with the original equalization payments (row 2 of Table 2) reproduced for convenient comparison in row 4. Not surprisingly, the row 5 equalization payments for all of the traditional five recipient provinces fall substantially while the payment for Ontario rises. In particular, Ontario's equalization rises from $249 per capita to $388 per capita. The decreases in equalization for Prince Edward Island, New Brunswick, and Manitoba are about $300 per capita while the decreases for Nova Scotia and Quebec are roughly $150 per capita.

While Ontario's share of overall equalization in 2012–13 from Table 1 was, as noted earlier, just over one-fifth, the $388 per capita value from row 5 of Table 3 would account for one-third of this same equalization total.[11] This further buttresses the conclusion that, under the assumption that the goal is to ensure that the recipient provinces can provide comparable *real* levels of provincial public goods and services, the current program over-equalizes the transfers to the traditional receiving provinces and under-equalizes the transfers to Ontario. Moreover, the amounts involved are anything but trivial.

However desirable it may be on economic or equity grounds to incorporate prices/costs into the calculation of equalization payments, this will obviously be difficult politically because it would significantly disadvantage six of the seven current have-not provinces. Nonetheless, some implications and recommendations will be proffered later. For present purposes, one technical recommendation is surely in order, namely, that Ottawa alter its approach when the formula-driven equalization exceeds the equalization cap. Currently, the approach is to reduce the per capita value of the equalization standard until overall payments fall within the cap. This reduces all provinces' per capita equalization by the same amount. But in the process the percentage reduction in Ontario's equalization is the largest (assuming that it remains the richest of the recipient provinces). However, given that Ontario is already considerably disadvantaged in terms of providing comparable real public goods and services, this approach exacerbates Ontario's disadvantage. An equi-proportional reduction seems more appropriate; that is, take the initial recipient provinces' percentages of overall formula-based equalization and then apply these same percentages to the allowable amount of equalization.

By way of a concluding comment, if the United States can fall back on the importance of capitalization as a rationale for not having an equalization program, then it seems inappropriate on Canada's part to completely ignore the role of costs and prices in our equalization program. This is even more the case since a straight-

---

[11] Ontario's share would be even larger absolutely and relatively if the row 5 exercise allocated equalization to achieve *real* per capita equality across the six recipient provinces. Rather, the corresponding deficiencies were filled with nominal dollars, not province-specific real dollars.

forward reading of section 36(2) would appear to support taking the prices/costs of producing provincial public goods and services into account.

However, even if Canada embraced the concept of incorporating prices into the definition of comparable levels of public goods and services, might it not be the case that, say, New Brunswick or Quebec would need a larger number of these comparable bundles? Readers will note that correcting for capitalization as a first step and then assessing the number of price-corrected bundles that different provinces may require is not the generally accepted approach to the concept of fiscal needs. Rather, the generally accepted approach of expenditure-needs advocates is that an expenditure-needs approach should simultaneously incorporate both differential prices/costs and differential physical needs or requirements.

I will defer my reflections on the appropriate approach until later. In the interim the analysis now turns to the most sophisticated assessment of the generally accepted vision of expenditure-needs equalization.

Expenditure-Needs Equalization: The Peter Gusen Analysis

Peter Gusen (2012a, 2012b) has recently, and courageously, undertaken an impressive and comprehensive approach to developing and measuring an expenditure-needs component of equalization, one that embraces both differential costs/ prices and measures of actual (price/cost independent) needs. In more detail, his results for expenditure needs in the various provinces are based on a weighted average over five expenditure areas – health care, elementary and secondary education, post-secondary education, social assistance, and "other social services." For these areas, he assesses the physical needs involved (e.g., the percentage of population of school age for elementary and secondary education needs, and the age distribution of the population for health needs). On the prices/costs side, given that the provincial price indexes in row 2 of Table 3 are also weighted averages of these five same expenditure categories, Gusen obtains overall expenditure needs by marrying the relevant prices/costs with the associated measures of physical need for each of the five categories.

The results for Gusen's approach for fiscal year 2008–09 appear in Table 4. The first five rows contain the estimates for relative provincial expenditure needs (i.e., deviations from the all-province average) for the five expenditure categories. The measures of overall expenditure needs by province (i.e., the sum of rows 1 through 5) appear as row 6. These measures of fiscal needs sum to zero across all provinces.

The results are probably not what one would have expected. Only three provinces have less than average overall needs – PEI, Manitoba, and Quebec, with the latter recording expenditure needs that are over $3 billion less than the all-province-average measure of needs.

Table 4: Equalization and Expenditure Need
(2008–09, $million)

	NL	PEI	NS	NB	QC	ON	MB	SK	AB	BC
1. Health	165	–19	167	165	–467	–178	–28	289	–848	752
2. Elementary and secondary education	–43	–9	–154	–96	–1,242	753	191	195	896	–491
3. Post-secondary education	–45	–39	16	–133	–783	286	–89	–70	1,141	–285
4. Social assistance	63	–15	46	89	675	26	–160	–82	–821	179
5. Other social services	170	–5	219	57	–1,300	–66	44	0	157	724
6. Total expenditure needs: sum of 1–5	310	–86	294	83	–3,117	822	–42	332	526	879
7. Fiscal equalization entitlements[ab] (2008–09)	0[c]	322	1,319	1,406	7,632	3,713	1,572	–352	–11,527	–1,379
8. Expenditure-needs equalization (row 6 + row 7)[b]	310[c]	235	1,613	1,489	4,515	4,534	1,530	0	0	0

Notes:

[a] Based on the equalization formula.

[b] Negative sums are set equal to zero.

[c] For this year NL was an equalization-receiving province according to the formula. But the formula includes only 50 percent of energy/resource royalties. There is a further provision, namely, that a recipient province cannot have an all-in revenue total (including 100 percent of energy royalties) that exceeds that for the lowest non-recipient province. Under this provision, an additional 50 percent for NL would exclude it from receiving equalization. Hence in the table, the NL figure in row 7 is set at zero. Indeed, with 100 percent inclusion of energy, the negative equalization entitlement for NL would exceed the $310 million in row 6 so that the corresponding entry for NL in row 8 under this scenario would be zero. This is my elaboration of Gusen's explanation for the NL data in rows 7 and 8.

Source: Gusen (2012a), Tables 11 and 12.

Row 7 contains the data on equalization payments for fiscal year 2008–09, reproduced from the Gusen paper. The final row of the table presents "expenditure-needs equalization," defined as the sum (where positive) of fiscal equalization transfers and the measures of expenditure needs. Where the sum of rows 6 and 7 is negative, the value in row 8 is set equal to zero. Gusen (2012a) assesses the row 8 figures as follows:

> While the change in overall payments [from row 7 to row 8] is relatively small, there would be a major shift in the provincial distribution of payments:
>   – Ontario and Quebec would receive equally large Equalization payments of just over $4.5 billion each. Under the current fiscal-capacity-only system [row 7], Quebec receives more than twice as much as Ontario, $7.6 billion versus $3.7 billion.
>   – Nova Scotia would enjoy a 20 per cent increase in its payments, from $1.3 billion to $1.6 billion.
>   – PEI would see a near 30 per cent drop, from $322 million to $235 million.
>   – Newfoundland and Labrador would become an Equalization recipient.
> The most significant conclusion of this exercise is probably the major redistribution of Equalization payments among provinces arising from expenditure need. It guarantees that discussions of expenditure need, or even the decision to put such discussions on the agenda, will be a contentious interprovincial issue. (27-28)

Although the expenditure-need figures sum to zero across all provinces, the equalization-receiving provinces receive roughly $1.7 billion less in positive entitlements than do the have provinces, with the result that overall equalization in row 8 is roughly $1.7 billion less than the row 7 total. While this could be adjusted to fit within an overall ceiling and/or floor, the appropriate nature of such an adjustment is not evident.

Capitalization and Needs: Some Personal Reflections

In principle, one would want to ensure that provinces faced with higher costs/prices for providing public goods and services would end up with higher per capita revenues, other things equal. Beyond this, provinces with greater physical needs (e.g., more elderly citizens per capita) should also require higher per capita revenues in order to address these needs, again other things being equal. In practice, however, addressing capitalization within equalization would seem to be on much firmer ground than addressing the physical needs component. To see this, it is instructive to note that were all national programs for Canadians designed so that similarly situated citizens were treated similarly no matter in which province they reside, then a stronger case could be made for incorporating an expenditure-needs component within the equalization formula. But this is clearly not the case. We have tended to take account of needs within many of the national programs. The privileged entry requirements and the more generous benefit-duration periods for high-unemployment regions within the EI program are cases in point. Relatedly, as reflected in the often-referenced table produced by Battle, Mendelsohn, and Torjman (2006), all cities west of Montreal had a much smaller percentage of their unemployed qualifying for EI than those east of (and including) Montreal. Also of relevance here is the large and long-standing divergence in federal immigration settlement funding that favours Quebec over Ontario. One presumes that these measures can be viewed as part of Canada's approach to addressing (or redressing) regional disparities. In reality they should be viewed as "equalizing components" on the needs front that are substitutes for provincial expenditures (e.g., for welfare payments in the case of EI). To the best of my knowledge, these types of equalizing components were not taken into account in the Gusen analysis so that the overall expenditure-needs equalization will tend to be overstated to this degree in Table 4 for most of the traditional recipient provinces.

There is another, more philosophical reason why I have a problem with embracing non-price-related needs. All of our federal-provincial transfers are unconditional, save for some national principles (e.g., no residency requirements for welfare, adherence with the Canada Health Act). Were we to engage in making transfers related to needs, pressures would develop to convert aspects of these transfers into conditional or specific-purpose transfers.

To see this, it is instructive to direct attention to Australia's comprehensive equalization program, which is operated by the Commonwealth Grants Commission (CGC). Many Canadian analysts look fondly on the Australian approach since it equalizes both revenue means and expenditure needs upwards and downwards to the national average. However, we usually do not recognize that the CGC allocations of the federally collected 10 percent GST to the Australian states represent only slightly more than half of the total transfers to the Australian states; the other half are Specific Purpose Payments (SPPs) that are conditional transfers. Not surprisingly, on occasion Australian lobby groups have put pressure on Canberra to convert parts of the CGC's unconditional grants transfers into SPPs, because some Australian states are not spending on the programs in question the amounts of money directed to these programs by the expenditure-needs component of the CGC grant system. The associated lobby typically argues that the particular need in question requires a Specific Purpose Payment in order to ensure that the money is appropriated to its intended purpose.

Since constitutionally and historically Canada is a much more decentralized federation on both the tax and the expenditure fronts than is Australia, one of the concerns with embracing expenditure needs in our equalization program is that this could become an avenue for reconditionalizing some of our unconditional transfers. There is a further problem with this possibility, namely, that the non-equalization-receiving provinces would not have their expenditures thus constrained.

In the final analysis, given the implications of the results in either Table 3 or Table 4, Gusen is probably right in that the capitalization and/or expenditure-needs options are not likely to see the light of legislative day. Nonetheless, I have two observations-cum-proposals that follow from the preceding analysis. The first is that since Quebec and Manitoba are much better off under the status quo than under the capitalization or needs exercises, it seems not that unreasonable to bring hydro rents more fully into the equalization calculations. The second is to note that probably the most important role we can assign to the implications arising from the Gusen exercise is to use them to forestall the introduction of measures that some provinces are currently harbouring, namely, embracing some specific program needs (e.g., percentage of elderly in the province) within equalization *irrespective of prices/ costs of providing these provincial public goods and services.*

Beyond this, the analysis to this point leads to an important policy recommendation. Given that the results in Table 3 and to a degree as well in Table 4 indicate that we are over-equalizing the traditional recipient provinces, there would seem to be a strong case for reworking our other national programs to ensure that similarly situated individuals, no matter where they reside, will be treated similarly. In other words, let us strip out the equalizing components embedded in other national programs and leave the task of equalization to the equalization program.

As a bridge between the above focus on the horizontal fiscal imbalances within the group of have-not provinces and the following section that addresses the vertical fiscal imbalances between the resource-rich provinces on the one hand and the non-resource-rich provinces on the other, it is tempting to contemplate a counterfactual (but not all that unrealistic) scenario where Ontario's fiscal capacity increases by just enough to make it a have province, all else remaining unchanged. Under this hypothetical scenario, nothing happens to Ontario's overall fiscal capacity since its equalization will fall apace. However, given that the total amount of equalization is fixed and Ontario's share has fallen to zero, the traditional recipient provinces will see their equalization (and their overall fiscal capacities) rise substantially. Although the current equalization formula does not allow for an equalization-receiving province to have a post-equalization fiscal capacity that exceeds that of the lowest have province, this is exactly what would happen in terms of purchasing power equivalent dollars. Indeed, and as noted earlier, this occurs already in that Prince Edward Island can provide more real bundles of provincial public goods and services than can have-province British Columbia (see row 3 of Table 3). The above counterfactual scenario would make this much more prevalent and ought to raise further concerns with respect to the likelihood of over-equalizing the traditional equalization recipients.

However, there is another challenging perspective that implicitly, at least, suggests that there is inadequate equalization across resource-rich and resource-poor provinces. To this the analysis now turns.

ENERGY ROYALTIES AND SURPLUS RECYCLING

Equalization in the Context of a Hydrocarbon and Hydroelectric Industrial Strategy

Canada's equalization program works tolerably well when provincial revenue sources are shared with Ottawa, as in the case for personal income taxes, corporate income taxes, and sales taxes for example. In part this is so because if, say, province A sees its corporate income tax (CIT) receipts rise significantly, which in turn will lead to an increase in overall equalization, then not only will Ottawa's CIT revenues also have risen apace (and therefore Ottawa can afford to pay the additional equalization) but there is a further appropriate coincidence, namely, that the province whose increase in corporate revenues has led to the increase in equalization is also the province whose residents (corporations) are conceptually funding the increase via their enhanced CIT payments to Ottawa's consolidated revenue fund.

On both counts, resource royalties are entirely different. First, thanks to sections 92(5), 92A, 109, and 125 of our Constitution, energy (and resource) royalties are constitutionally the prerogative of the provinces; they cannot be accessed by Ottawa. More generally, section 92A grants the provinces the right to raise money by any mode or system of taxation in relation to resource revenues. To my knowledge, Canada's treatment of royalties is unique among the world's federations: no other federation has anywhere near such a powerful provincial-rights provision. Second, and consequentially, the "who benefits/who pays" coincidence noted above is severed. Hence when energy royalties in Saskatchewan, Newfoundland, or Alberta increase, Ottawa has to finance the resulting increase in equalization from its consolidated revenue fund. which as noted, is not able to share in provincial energy royalties.[12] This means that if the province in question is Alberta, for

---

[12] Note that this is not to say that Ottawa does not benefit revenue-wise from the energy sector. It obviously does: on a per capita basis Alberta is the leading contributor to Ottawa's finances, and energy is clearly a major driver of economic activity beyond the borders of royalty-receiving provinces. Moreover, the CIT and other revenues arising from the energy sector fall in the earlier category – they are shared with Ottawa so that the resulting increase in equalization arising from, say, the energy-related CIT is financed from the corporate profits in the resource provinces. The point at issue here is that royalties are in a class by themselves. Indeed, around the time of the advent of the 1982 Fiscal Arrangement negotiations there emerged a literature arguing that the value of royalties that were spent on the production of provincial goods and services should be treated as an imputed benefit to residents for purposes of income taxation. The argument then was that this category of provincial public goods is the only type produced with monies that had escaped federal taxation. Note that

example, then its residents contribute somewhat more than their 11 percent population share (because they are a have province) to the financing, and Ontarians probably contribute somewhat less than their 39 percent population share. In other words, Ontarians will be saddled with close to two-fifths of the financing of an increase in equalization triggered by an increase in energy royalties in the resource-rich provinces.

Overall energy royalties accruing to the provinces in 2012–13 were $22.563 billion and are the principal reason why Newfoundland, Saskatchewan, and Alberta have overall per capita fiscal capacities well in excess of the other provinces – see Table 2. Our existing approach to accommodating these royalties within the equalization program is (1) to allow only 50 percent of these royalties to enter the program, and (2) because these royalties are the principal driver of the recent increases in equalization entitlements, to limit the annual growth of equalization payments to the rate of the three-year average of GDP growth. However, the reality is that these royalties may well loom even larger in years ahead. For example, the Canadian Energy Research Institute (CERI) estimates that the oil sands in Alberta will lead to $350 billion in provincial royalties over the next 25 years as well as $122 billion in provincial and municipal tax revenues.[13] To be sure, overall oil-sands-related Canadian tax revenue (excluding royalties) will increase by $444 billion with 70 percent ($322 billion) flowing to Ottawa.

These data relate only to energy tax revenues/royalties, not the level of economic activity, and then only to Alberta. In other words, the seemingly endless demand for our resources associated with the economic ascendancy of populous China and India is a game-changer, a potential economic bonanza that will only intensify as overall domestic and global economic activity recovers and as China and India begin to narrow the still-dramatic per capita income gap between themselves and the rich nations. The emerging response from influential policy leaders such as James Prentice (then senior vice-president and vice-chairman of CIBC and now premier of Alberta) is in the direction of a resource-based economic future, or what might be termed a hydrocarbon and hydroelectric industrial strategy. Prentice (2011) refers to this resource-based strategy as an "energizing infrastructure" opportunity as part of "Canada's 21st century nation-building." This twinning of fossil energy with hydroelectricity would bring Manitoba and Quebec under the industrial strategy umbrella (joining the three westernmost provinces and Newfoundland and Labrador). Moreover, by integrating hydro power with the less environmentally

---

this would also apply to royalties from hydroelectric power. This issue will not be pursued further in this chapter, although it may emerge again in the academic literature if hydroelectric royalties become increasingly important.

[13] CERI estimates are sourced from the Government of Alberta web entry http://oilsands. alberta.ca/economicinvestment.html.

benign oil sands and by developing a corresponding green energy policy, the overall energy strategy would arguably be made more saleable both at home and abroad.

Energy Royalties and Differential Provincial Fiscal and Economic Fortunes

The Achilles heel of such a hydrocarbon/hydroelectric strategy may well lie on the fiscal and federal (indeed fiscal-federalism) fronts. Or in terms of the theme of this chapter, the failure to find ways to (indirectly) recycle the resulting fiscal surpluses, interprovincially and federal-provincially, could seriously complicate, even undermine, any national resource-based industrial strategy.[14] There are two seemingly unrelated, but actually closely intertwined, issues at play here.

The first and most obvious is that a ratcheting up of resource royalties would dramatically increase the fiscal disparity between the resource-rich and the equalization-receiving provinces. This is because, as noted above, the federal government cannot, constitutionally, *directly* access provincial royalties, and so the prospect of tax havens and/or superior provincial public goods and services in resource-rich provinces becomes a distinct possibility. Hence, Ottawa has to find indirect ways of recycling these resource revenues, which is in large measure the subject matter of this section. By way of an instructive aside in relation to the tax-haven issue, Canadians ought to be thankful that Albertans abhor sales taxes. This is the most benign form of tax to eliminate because it has little impact on interprovincial factor flows. In sharp contrast, the interprovincial factor flows (including movement of corporate headquarters) would probably be quite dramatic were Alberta to have reduced its corporate income to zero rather than forgoing a sales tax. Moreover, a zero corporate tax would cost *less* in terms of forgone revenues than does a zero provincial sales tax (i.e., roughly $4 billion for the CIT vs. the earlier noted $6 billion for a provincial sales tax, both evaluated at national-average tax rates). Presumably Alberta recognized that Ottawa would probably have had to respond in a countering fashion to a zero CIT, so this may have also served to tilt the Alberta government's preference in favour of sales tax relief.

The second issue falls under the general rubric of the "Dutch disease," so named because Holland's exports of North Sea oil and gas appreciated its exchange rate to such a degree that this clobbered its manufacturing sector.[15] Given the utter volatility of the energy prices, my presumption relating to the Canadian reality is

---

[14] A more comprehensive approach to the surplus-recycling challenge of an energy industrial strategy would embrace the demands/rights of the First Nations.

[15] The term Dutch disease was coined by *The Economist* in 1977 to describe the decline of the manufacturing sector in the Netherlands after the discovery of a large natural gas field in 1959.

that our currency area is too small to accommodate at the same time a world class manufacturing sector and a global energy powerhouse. This is clear from Figure 1, which plots the rise in energy prices (in US dollars per barrel) and the value of the loonie (in US cents per Canadian dollar). The relationship is readily apparent: a rise in global energy prices generates export-driven resource income from, as well as inward foreign direct investment into, our energy patch, both of which will drive up the value of (i.e., appreciate) the loonie. In the process, the global price of resources rises relative to the global price of manufactures – a relative price change that will carry over to Canada. However, because resources play a larger role in the Canadian economy than they do in the US economy, the Canadian dollar will appreciate relative to the US dollar. But the near-doubling of the loonie in Figure 1 (from 62 US cents in 2002 to just over 110 cents in 2008) represents very significant *exchange-rate overshooting,* well beyond the appreciation required to accommodate the increase in resource prices relative to the price of manufacturers. Although not shown in Figure 1, there was an earlier and equally rapid depreciation in the 1990s that also represented exchange-rate overshooting, this time on the downward side.

Figure 1: US-Canada Exchange Rate and Crude-Oil Price, 2002Q1–2011Q4

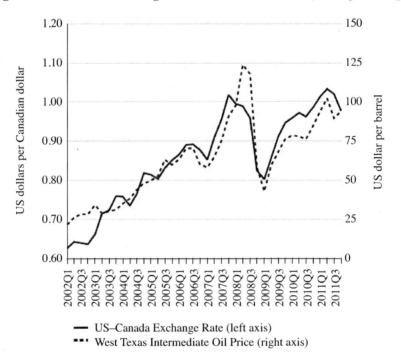

— US–Canada Exchange Rate (left axis)
■■■ West Texas Intermediate Oil Price (right axis)

Source: Bank of Canada; US Federal Reserve Bank of St. Louis, FRED data retrieval system. Reproduced from Courchene (2012, Figure 2).

To be sure, much more than the exchange rate was and is at work in terms of the sharp decline in our manufacturing sector. Specifically, given that the major markets for Canada's manufacturing are US consumers and manufacturers, the wholesale offshoring and outsourcing of US manufacturing to China in order to take advantage of the inexpensive but efficient Chinese labour force clearly played the dominant role in the shrinking of the Canadian manufacturing sector.[16] Nonetheless, and in contrast to the prevailing wisdom, my view has long been that the Bank of Canada should not have permitted swings in the loonie of anywhere near the magnitudes experienced recently. Indeed, even the Swiss monetary authorities, long viewed as the gold standard in the pantheon of central bankers, are now intervening in currency markets to limit the appreciation of the fabled Swiss franc relative to the euro.

Prior to turning attention to these indirect approaches to surplus recycling, it should be noted that there is one option that would qualify as a *direct* surplus-recycling mechanism, namely, a direct transfer of royalties from one province to another. Not surprisingly, this option has arisen in the BC-Alberta stand-off over the Enbridge Northern Gateway pipeline. Given that Alberta will pocket scores of billions in royalties while British Columbia will be saddled with any environmental catastrophe, it is only natural that British Columbia would be interested in securing a share of Alberta's benefits and/or adequate compensation for any environmental disaster as quid quo pro for the pipeline to proceed, all other factors being onside. This is an obvious example of the importance of having adequate surplus-recycling mechanisms in place in order to pave the way for the emergence of a comprehensive hydrocarbon and hydroelectric industrial strategy.

Focus is now directed to other options, not all in conflict with the interests of the resource-rich provinces, which can serve the same purpose.

Options for Ameliorating Resource-Driven Interprovincial Fiscal Imbalances and the Dutch Disease

Stewardship as a Principled Perspective

In his insightful June 2012 *Policy Options* article "Reversing the Curse: Starting with Energy," David Emerson provides a principled perspective for addressing the

---

[16] As noted in the introduction to this chapter, US manufacturing has also been harmed by something akin to the Dutch disease. The trigger for offshoring to China was the desire to take advantage of the cheap but efficient Chinese labour force. However, because the Chinese sterilized the inflow of dollars, the US dollar remained overvalued relative to the yuan, and the downward manufacturing spiral continued. The recent shift toward "re-shoring" back to the United States owes more to the dramatic rise in Chinese wages than to any appreciation of the yuan.

implications of resource revenues for both the Dutch disease and interprovincial fiscal equity. This principled perspective is *stewardship*:

> A longer-term, more disciplined approach to managing energy and resources is required. Natural resources are long-term assets that belong to generations of Canadians now and into the future. Government leaders and decision-makers have an implied custodial and stewardship responsibility to manage across the generations. In fiscal and economic terms, non-renewable energy and natural resources are long-life, fixed assets that, when sold and monetized, should be reinvested in ways that will benefit Canadians over the long term. *Pretending that resource revenue is just another form of operating revenue, to be spent on current consumption of public services, is an abrogation of this responsibility.* (53, emphasis added)

In the economics literature, the optimal approach to non-renewable-resource stewardship is the "Hartwick Rule" (named after my Queen's colleague John Hartwick); namely, non-renewable assets when sold should be invested, and the annual return on this investment can be spent or, in the context of this chapter, bought into provincial budgets.

Attention is now directed to alternative ways in which resource revenues can be indirectly recycled. Much of what follows has its roots in the existing Canadian policy literature. The most recent contributions would include Boadway, Coulombe, and Tremblay (2012), Tremblay (2012), and Courchene (2012).

Provincial Sovereign Wealth Funds (PSWFs)

This stewardship perspective points in the direction of PSWFs, preferably along the lines of Norway's sovereign wealth fund. Fuelled by fossil energy revenues, Norway's fund is invested in international markets. This serves to offset Norway's energy-related export earnings, thus in turn serving to ameliorate the tendency for the Norwegian currency (krone) to appreciate. PSWFs invested in international markets would play the same role – stewarding energy-related revenues for use by future generations and in the process reducing the magnitude of the Dutch disease. As noted earlier, reducing the degree to which the loonie would appreciate in the face of an increase in the international demand for and/or price of energy would mean more Canadian dollars for any given level of exports, a clear gain for both energy exporters and governments alike.

Were Alberta to have introduced a sales tax and created a PSWF (or continued with the Heritage Fund), the current value of such a fund would be well in the hundreds of billions of dollars. Indeed, a further role for such a fund could be to stabilize the province's overall revenues in the face of either revenue shortfalls or excesses. To be sure, a PSWF in the hundred-billion-plus area would likely create a challenge of its own to the federation.

Since the energy revenues placed in PSWFs would not enter provincial consolidated revenues for budgetary purposes and, therefore, would not be devoted to the provision of current public goods and services, *these revenues should not enter the equalization formula.* However, when funds are withdrawn from PSWFs and brought back into provincial consolidated revenue funds, they would then enter the equalization formula.

Other Avenues for Sheltering Royalties from the Equalization Program

As the Emerson quotation notes, investing royalties in ways that will benefit Canadians over the longer term is also consistent with resource stewardship. An example might be paying down provincial debt. On the one hand such funds would not (or should not) enter the equalization formula since they are not being used to provide current goods and services and, on the other, they are benefiting citizens over the long term. Selected capital expenditures would also qualify under this rubric.

Redesigning Federal Corporate Profits Taxes

While Ottawa cannot directly access provincial royalties, it can alter its corporate taxation of the energy sector in ways that will increase its revenues from the sector. The obvious, albeit controversial, approach here would be to disallow deduction of a corporation's royalty payments to provincial governments in calculating its federal corporate taxes. One likely result of this would be that the provinces would be put under pressure to reduce their royalty rates. On the other hand, it is the ability of energy firms to deduct provincial royalties in calculating federal corporate taxes that allowed the provinces to set higher royalties in the first place.

An increasingly appealing alternative, in part because it is becoming more acceptable internationally, would be to convert the corporate tax system into a tax on rents. Boadway, Coulombe, and Tremblay (2012) reflect on this proposal as follows:

> A tax on rents would capture revenues for the public sector from rents or pure profits generated from all sources, including monopoly rents, resource rents, locational rents, and rents due to special advantages. A corporate tax based on rents would generate for the federal government a share of resource rents using a tax that is not explicitly discriminatory, and would contribute to the federal government's ability to address fiscal imbalances arising from natural resources.

Revenue Testing Federal-Provincial Transfers

Canada income tests virtually all its transfers – Guaranteed Income Supplement, Old Age Security, Employment Insurance, Canada child tax benefits, welfare benefits, and probably others. The time has come to "revenue test" the equal per capita federal-provincial transfers to the provinces. In an earlier article (2010), I proposed that the CHT/CST combination be subject to revenue testing along the following lines. Using the all-in fiscal capacity as measured by row 5 of Table 2, if a province has a per capita all-in fiscal capacity above a certain threshold, say 115 percent, of the per capita national average of all-in fiscal capacities, then for each dollar per capita of a province's revenues above this threshold, Ottawa would reduce its CHT/CST transfer by, say, 25 cents per capita. Given that the current value of the CHT/CST is roughly $1,200 per capita (row 6 of Table 2 above), if a province has an all-in fiscal capacity of $4,800 per capita above the 115 percent per capita threshold, then its CHT/CST will fall to zero. The resulting CHT/CST clawbacks could then be redistributed to the provinces with per capita revenues below the threshold.

A few comments are in order. While provincial fiscal capacity for purposes of this recommendation would include all revenues, it is likely that energy royalties will be the primary reason for per capita provincial revenues in excess of 115 percent of the national average. However, this does not amount to a confiscation of royalties any more than a reduction in one's old age pension due to an increase in earned income amounts to a confiscation of the earned income. Moreover, unlike the 100 percent clawbacks on the Guaranteed Income Supplement (GIS), a 25 percent revenue clawback is rather moderate. Indeed, under the Australian Commonwealth Grants Commission approach, the clawback of revenues (say for Western Australia's large resource-related revenues) is effectively 100 percent once they exceed the all-state per capita average.

One of the hallmarks of our approach to the social envelope in comparison with the United States is that we engage in targeting-cum-income-testing for virtually all of our social benefits whereas the Americans do not; their social security payments are universal rather than targeted (via income testing) to those most in need. In other words we purchase more equity, as it were, than do the Americans from every dollar of social policy spending. Revenue testing is a natural extension of income testing.

It should be clear that there is nothing sacrosanct about the choice of the 115 percent threshold or the 25 percent clawback. Others would probably choose different parameters. But what hopefully becomes acceptable is that revenue testing, already the cornerstone of our equalization program, also becomes a defining feature of the rest of our federal-provincial transfer system.

Finally, and as noted in the context of the discussion of Table 2, it is instructive to recognize that *the CHT/CST has been subject to revenue testing*. The precise details are arcane but, in general terms, provinces with high per capita revenues from the personal tax and to a lesser degree the corporate income tax received smaller per capita CHT/CST cash transfers. The 2007–08 federal budget committed Ottawa to ensuring that the CHT/CST would henceforth be equal per capita across the board for all provinces. Ontario has now been brought up to the other provinces' level, and Alberta will get there in fiscal year 2014–15. The message here is twofold: (1) revenue testing the federal-provincial cash transfers is not new, and (2) meaningful indirect surplus recycling requires that it be reinstated along the lines outlined above.

Pricing Carbon Emissions

The fiscal federalism issues associated with carbon pricing provide a convenient transition between the current issues relating to the possibility of resource-rich provinces morphing into tax havens and/or suppliers of superior public goods on the one hand and the core issue in the final substantive section dealing with the likelihood that some provincial governments will have inadequate revenue sources to meet their growing expenditure responsibilities on the other. Given that the potential revenues arising from carbon pricing could be very substantial, were the bulk of these revenues from pollution abatement to accrue, via upstream or origin-based emission taxes, to the energy-rich provinces that are already receiving huge energy rents/royalties, then this would dramatically exacerbate the already challenging differential fiscal capacities across provinces. In a *Policy Options* article, John Allan and I (March 2008) argued that the preferred option would be a nationally run, destination-based (i.e., a final-consumption-based) carbon tax regime. Among the reasons for this were that (1) the burden of CO_2 affects all Canadians more or less equally; (2) the provinces cannot prevent "carbon leakage" because they cannot levy tariffs interprovincially or internationally on products produced under less stringent carbon-pricing regimes whereas Ottawa can; and (3) while some of the revenues should be devoted to R&D related to developing low-carbon technologies and processes, the rest of the revenues collected could be distributed to the provinces on an equal per capita basis. Allan and I also recommended that Ottawa should treat carbon taxation as it relates to international trade along the lines of the GST or value-added taxation, namely, apply the carbon taxes to imports and provide carbon-tax rebates on exports.

Under such a scheme, carbon taxation would be export-import neutral, would stimulate low-carbon technologies, and would allocate the very substantial carbon-abatement revenues equally in per capita terms across provinces, thereby serving not only to ameliorate the existing interprovincial fiscal capacity differentials but also to addressing the looming imbalance in the division of money and power between Ottawa and the provinces.

CONCLUSION

The conclusion of this chapter is as straightforward as it is important. If effective surplus-recycling systems are essential to the stability and resilience of macroeconomic systems (including federations), as I believe they are, then the reality that Canada's surplus-recycling systems as they relate to interprovincial fiscal imbalance are far from effective ought to be of major concern alike to our political leaders and to the Canadian policy community.

REFERENCES

Battle, Ken, Michael Mendelsohn, and Sherri Torjman. 2006. "Towards a New Architecture for Canada's Adult Benefits." Ottawa: Caledon Institute of Social Policy.

Boadway, Robin, Serge Coulombe, and Jean-François Tremblay. 2012. "The Dutch Disease and the Canadian Economy: Challenges for Policy Makers." Paper prepared for Thinking Outside the Box: A Conference in Celebration of Thomas J. Courchene, Kingston, October 26–27.

Courchene, Thomas J. 2004. "Confiscatory Equalization: The Intriguing Case of Saskatchewan's Vanishing Energy Revenues." *Choices* 10 (2). IRPP, Montreal.

—. 2008. "Fiscalamity! Ontario: From Heartland to Have-Not." *Policy Options/Options politiques* (June): 48-54.

—. 2010. "Intergovernmental Transfers and Canadian Values: Retrospect and Prospect." *Policy Options/Options politiques* (May): 32, 40.

—. 2011. *Rekindling the American Dream: A Northern Perspective*. The 2011 IRPP Policy Horizons Essay.

—. 2012. *Policy Signposts in Postwar Canada: Reflections of a Market Populist*. Montreal: Institute for Research on Public Policy.

Courchene, Thomas J., and John R. Allan. 2008. "Climate Change: The Case for a Carbon Tax." *Policy Options/Options politiques* (March): 59-64.

Department of Finance Canada. 2013. "Federal Support to Provinces and Territories." http://fin.gc.ca/fedprov/mtp-eng.asp.

Emerson, David L. 2012. "Reversing the Curse: Starting with Energy." *Policy Options/Options politiques* (February): 53-57.

Gerson, Jen. 2012. "Borrowing the Blues: Alberta's Dance with Debt May Put on a Slippery Slope." *Financial Post*, November 19, A1 and A6.

Gusen, Peter. 2012a. *Expenditure Need: Equalization's Other Half*. Toronto: Mowat Centre for Policy Innovation.

—. 2012b. *Expenditure Need: Equalization's Other Half: Technical Report*. Toronto: Mowat Centre for Policy Innovation.

Oates, Wallace. 1983. *Fiscal Federalism*. New York: Harcourt Brace Jovanovich.

Prentice, Honourable Jim. 2011. "Nation-Building Infrastructure: A Roadmap to Economic Growth." Address to the Edmonton Chamber of Commerce by CIBC's senior vice-president and chairman, November 21.

Tremblay, Jean-François. 2012. *Fiscal Problems, Taxation Solutions: Options for Reforming Canada's Tax and Transfer System*. Toronto: Mowat Centre for Policy Innovation.

Varoufakis, Yanis. 2011. *The Global Minotaur: America, the True Origins of the Financial Crisis and the Future of the World Economy*. London and New York: Zed Books.

6

EQUALIZATION AND THE POLITICS OF NATURAL RESOURCES: BALANCING PROVINCIAL AUTONOMY AND TERRITORIAL SOLIDARITY

Daniel Béland and André Lecours

INTRODUCTION

Equalization, the so-called "glue that holds the federation together" (Bryden 2009, 76), often seems to be anything but. In fact, the equalization program, a central element of fiscal federalism and territorial redistribution in Canada, sometimes generates intergovernmental conflict that seems antithetical to the national social cohesion the program was intended to promote (Lecours and Béland 2010). All federal systems represent balancing acts between territorial (substate) autonomy and solidarity, a sociological concept that is not necessarily used by citizens and policy actors but that stresses a crucial institutional and political imperative ever present within these systems. In a country like Canada, where provincial autonomy is politically and constitutionally entrenched and supported by strong territorial (provincial) identities, the challenge to mechanisms of territorial redistribution and solidarity is considerable. Therefore, it comes as no surprise that, from time to

The authors thank Loleen Berdahl, Tom Courchene, Frank Graves, André Juneau, and the other participants of the 2012 State of the Federation conference for their comments and suggestions. Daniel Béland acknowledges support from the Canada Research Chairs Program.

time, provincial governments think that they are getting "shafted"[1] by the federal equalization program, either because they feel that they do not receive the payments they deserve or because, as non-recipient provinces, they believe that they have nothing to gain from the program.[2]

In Canada, intergovernmental conflict over equalization has several different sources, but a quick survey of the debates shows that a specific one stands out: the uneven territorial distribution of natural resources (especially oil and natural gas) across the country. In the context of high resource prices and revenues and the structural decline of the Ontario manufacturing sector, the political economy of the country is changing in a way that puts pressure on the federal equalization program, which is now allocating money to Ontario for the first time since its inception in 1957. Shifting patterns of territorial economic inequalities are creating discontent in provinces seeking to develop their non-renewable natural resources (for example, Saskatchewan and Newfoundland and Labrador). Such a reality also increases disparities linked to rising oil and gas prices while exacerbating cynicism toward the equalization program in Alberta. Provinces with no significant oil and gas resources are struggling to keep pace with their (non-renewable) resource-rich counterparts, and are likely to remain in the equalization recipient category for some time.

This chapter examines the changing dynamics between equalization and natural resources by focusing on the political tensions between territorial solidarity and provincial autonomy. The chapter is divided into three main sections. In the first section, we discuss how the creation and development of the federal equalization program was grounded in a logic of territorial solidarity. In the second section, we suggest that, in the historical and institutional context of Canadian federalism, natural resources are closely tied to the idea of provincial autonomy.[3] In the third section, we explore the politics of collision between the logic of territorial solidarity (inherent to equalization) and the idea of provincial autonomy (associated with natural resources).

---

[1] This was the terminology used in 2007 by Newfoundland and Labrador premier Danny Williams (CBC News 2007).

[2] One could argue that even non-recipient provinces have benefited from equalization because, without it, the Canadian tax system would probably have been centralized as poorer provinces would have not accepted such a high level of institutional fragmentation. The authors wish to thank Tom Courchene for his insight on this issue.

[3] For a similar argument focusing on the West, see Janigan (2012).

EQUALIZATION: FOSTERING UNITY AND SOLIDARITY IN CANADIAN FEDERALISM

Federal systems often represent answers to governance issues stemming from ethno-linguistic diversity (Canada, India, Switzerland), from a history of territorial autonomy in the context of state formation (the United States and Australia), and from the administration of a large territory (Brazil and Russia). The federal answer to these challenges rests in the principle and mechanisms of decentralization, which allow for the autonomous governance of communities who desire it and offer potential for the effective control of territory. An almost unavoidable pitfall of decentralized arrangements is the potential for citizenship to lose some of the integrative and unifying power it tends to hold in more centralized liberal-democratic contexts, as the existence of multiple, autonomous, and democratically elected governments within the same polity helps produce and reproduce political communities with their own identities, political elites, and institutions. The political identities stemming from federalism need not be in competition with the (state) national identity, but the potential for this type of dynamic is there. Moreover, as a result of political decentralization, federal systems run the risk of having variable quality in public services across constituent units, as some of these units will almost unavoidably have greater fiscal means than others. To address this situation and, more generally, to build national unity, federal governments typically make equalization payments to constituent units that find themselves below a certain (countrywide) fiscal standard. Not only do these payments help citizens gain access to public services of comparable quality across the land, but they represent concrete expressions of abstract notions of national unity and provide poorer jurisdictions with incentives to stay within the existing federal system. This is especially true for multinational federations (Burgess and Pinder 2007), where independence may already be a genuine political alternative.

Indeed, equalization payments can make federalism seem like a desirable fiscal proposition for a minority group living within a lesser-off constituent unit. Hence, even those within this minority group who might feel that their constituent unit is politically marginalized within the federation, or that the federal government promotes and projects an identity that is not theirs, might accept the political status quo. Equalization does not only work toward unity in federal states by "buying off" poorer constituent units; it can, in the longer term, work to create a community of redistribution that may generate feelings of countrywide togetherness and solidarity.

National unity, which is directly tied to the issue of territorial solidarity, was of foremost concern to the federal politicians and bureaucrats who created the equalization program in 1957. Nearly two decades earlier, in its 1940 report, the Rowell-Sirois commission had suggested the creation of National Adjustment Grants to be paid to the provinces on the basis of fiscal need (Courchene 1984,

65). According to the report, these grants would foster national unity by helping to prevent citizens from poorer provinces from feeling neglected by the rest of the country (Royal Commission on Dominion-Provincial Relations 1940, 79). A decade or so later, the focus became the better integration of Quebec within the federation. After 1945, Quebec spearheaded a challenge to the tax rental system that had been created during the war, and the province opted out of the system in 1947. For Quebec's Union Nationale government, the tax rental system acted as an intrusive and centralizing scheme because it involved Ottawa taking over many provincial taxes in exchange for federal transfers to the provinces.

In 1954, the Union Nationale went further and reimposed the provincial income tax. That same year, the province's Tremblay Report supported autonomist positions, especially on fiscal federalism. These key developments, which were widely interpreted as challenges to the existing structures of the Canadian federation (Milne 1998, 190), were central to the creation of the federal equalization program in 1957 (Bryden 2009; Courchene 2012, 4-5). From that moment, the allocation of "equalization payments, regardless of whether or not a particular province rented its tax fields to the national government, would be a way of ending the isolation of Quebec" (Bryden 2009, 81). The creation of an equalization program was also supported by the Ontario government, which felt that such an initiative could serve to bring poorer provinces like New Brunswick on board in the fight against fiscal centralization (Bryden 2009). Indeed, Ontario and Quebec were then leading the charge against federal power, as exemplified by the dynamics of the two (1945 and 1946) dominion-provincial conferences on reconstruction, during which Ontario premier George Drew forged an alliance with Quebec premier Duplessis to openly oppose federal plans for the centralization of the country's fiscal arrangements. For the federal government, the fact that Quebec opted for its own income tax led to a concern that other provinces would do the same. In a Keynesian context, Ottawa wanted to keep control of a macroeconomic instrument such as income tax, and equalization facilitated this objective.

The logic of solidarity, both socioeconomic and political, was thus foremost in the creation of the federal equalization program. Provinces that have consistently qualified as recipients have steadfastly supported the program. Indeed, in Prince Edward Island, New Brunswick, Nova Scotia, and Manitoba, equalization payments currently account for anywhere between 10 and 20 percent of provincial budgets. In these provinces, such payments represent concrete manifestations of Canadian solidarity, as they are instrumental in enabling provincial governments to provide public services of a quality comparable to those offered in wealthier provinces. Even in Quebec, equalization receives solid backing and has served as an instrument of nationalist accommodation. Quebec politicians who oppose independence frequently argue that Canadian federalism is a worthwhile financial proposition for their province. Long-time Quebec premier Robert Bourassa famously spoke of *le fédéralisme rentable* (profitable federalism) to highlight the concrete benefits for

Quebecers of staying within the Canadian federation. Equalization payments are a major component of this *fédéralisme rentable*.

For Quebec Liberals (PLQ), equalization is part of Canadian citizenship. They simply argue that it is the right of Quebecers as Canadian citizens to have their province receive equalization payments from Ottawa.[4] Liberals explain that, although Quebec's position as a recipient province is an unfortunate situation, it is the product of both its late industrial development and its lack of non-renewable natural resources such as oil and gas. Consequently, there is no shame in receiving equalization payments. In fact, former Quebec Liberal premier Jean Charest (2003–2012) even suggested that the program should be enhanced, finding that the unconditional nature of equalization payments makes it a better funding structure than conditional transfers (Séguin 2004).

For sovereignist politicians, equalization represents a political problem. Their argument for an independent Quebec means that they can never ascribe any value to Canadian citizenship. As such, they typically want to avoid speaking about equalization and, when forced to do so, will attempt to put it in a broader fiscal and policy context.[5] Sovereignists suggest that, in the overall scheme of fiscal federalism, Quebec does not gain anything; in fact, they argue, it probably comes out a loser, in part because they claim the federal government has invested more heavily in other provinces than in theirs. Therefore, the sovereignist view on equalization is that it simply "sends money back" to Quebecers. This position has received some support in Quebec. For example, in a 2009 public opinion poll 31 percent of Quebecers said that they paid more income tax to the federal government than the federal government spent in the province, against 23 percent who said the opposite (30 percent felt things evened out, while 16 percent did not know).[6]

There is a third perspective on equalization in Quebec that implicitly endorses the notion that Quebecers as Canadian citizens have a right to equalization payments, but characterizes as shameful the province's status as a recipient. The defunct Action démocratique du Québec first articulated this position, which was picked up by Coalition Avenir Québec (CAQ). In the 2012 Quebec election campaign, CAQ leader François Legault even said that his objective was that "in 10 years, Quebec will pay equalization to the rest of Canada" (Leblanc 2012).

Beyond Quebec, equalization receives strong support in the other traditional recipient provinces: Manitoba, New Brunswick, Nova Scotia, and Prince Edward Island. This is hardly surprising, as these provinces' budgets depend quite substantially on equalization (close to 20 percent in the case of PEI), although Quebec receives the largest share of equalization payments as a result of its much larger

---

[4] Interview with a Quebec government equalization analyst, January 2011.

[5] Interview with a Bloc Québécois (BQ) member of parliament, November 2010.

[6] CROP survey, April 2009, http://ideefederale.ca/wp/?p=7.

population. In Manitoba, New Brunswick, Nova Scotia, and PEI, equalization is therefore seen as a key source of fiscal solidarity that renders available quality public services in spite of a lower than average fiscal capacity. This attachment also comes with expectations concerning the predictability and stability of equalization payments. Just as large health and social programs such as medicare create large constituencies who are likely to fight for the preservation of these programs (Pierson 1996), the equalization program has generated powerful vested fiscal and political interests in receiving provinces.

The solidarity imperative embedded in the equalization program acquired a new status with the constitutionalization of equalization in 1982.[7] By stating that "Parliament and the government of Canada are committed to the principle of making equalization payments to ensure that provincial governments have sufficient revenues to provide reasonably comparable levels of public services at reasonably comparable levels of taxation," the Constitution Act of 1982 (section 36.2) clearly makes equalization, and the territorial distribution it involves, a major component of the Canadian polity. Canada is, by virtually all indicators, one of the most decentralized federations in the world, which means the central government has, comparatively speaking, relatively few instruments to advance the solidarity, or the shared-rule, dimension of federalism and to enact concrete programs embodying such solidarity. Equalization is one of these instruments.

NATURAL RESOURCES AND PROVINCIAL AUTONOMY

Canada is often said to be a resource economy, which typically does well when commodity prices are high. There is, however, a territorial dimension to this picture insofar as resources, especially much sought after non-renewable resources such as oil and gas, are unevenly distributed among provinces. Consequently, natural resources are a key source of differential fiscal revenues among provincial governments, as some governments depend much more on such revenues than others. Indeed, in the context of the decline in the Ontario manufacturing sector, provincial fiscal capacity in Canada mirrors the distribution of non-renewable resources.

This being said, there is more to natural resources in Canada than simply their economic value. Indeed, resources are, in many provinces, central to provincial identity and the idea of provincial autonomy (for instance, see Janigan 2012). As such, natural resources factor heavily into Canadian federalism, intergovernmental relations, and politics. The connection between natural resources and provincial autonomy has a strong constitutional basis: the 1867 British North America Act

---

[7] A provision about equalization was included in the original constitutional Accord and, during the deliberation process, no serious opposition to its inclusion in the 1982 Constitution Act emerged. We would like to thank Roy Romanow for his insight on this issue.

assigns to provinces responsibility over "the Management and Sale of the Public Lands belonging to the Province and of the Timber and Wood thereon" (92.5).[8] Furthermore, the Constitution Act of 1982 specifies that the provinces have the exclusive powers to make laws on the exploration, management development, and conservation of non-renewable natural resources, forestry resources, and electrical energy (92A). When British Columbia and Prince Edward Island joined the federation in 1871 and 1873, respectively, their natural resource ownership was fully recognized, just as it had been for the first four provinces (New Brunswick, Nova Scotia, Ontario, and Quebec). The situation was different in the Prairies. In 1870, arguing that it needed to control western resources to offset the costs of railroad construction, Ottawa denied resource ownership to Manitoba when it joined the federation. In 1905, Alberta and Saskatchewan received the same unfavourable treatment (Thompson, n.d.). After a long political battle with Ottawa, the three Prairie provinces finally gained full control over their natural resources in 1930. "To mark the handover of resource control, he [Prime Minister Mackenzie King] presented a cheque for \$4,822,842.73 to [Manitoba Premier] Bracken" (Janigan 2012, 328-29). Despite this, the treatment of resources within Canadian federalism has remained a sensitive question in Western Canada.

In Alberta, oil and gas define much of the province's economy and politics (Tupper, Pratt, and Urquhart 1992, 35-36). Alberta's heavy reliance on non-renewable natural resources for its economic development exacerbates political anxieties grounded in a sense of institutional vulnerability vis-à-vis Ottawa. So-called Western alienation, as it pertains to Alberta, stems primarily from the fact that the province has never been well represented within the Liberal governments that have held power in Ottawa for most of Canada's history, but it has also been fuelled by the sense that the federal government has syphoned provincial wealth, which is mainly derived from resources (agriculture in the early years, oil and gas in more recent decades), to benefit other parts of the country. In Alberta, the Trudeau government's 1980 National Energy Policy (NEP), enacted in the context of worldwide concerns about oil availability and affordability, magnified political discontent toward Ottawa while galvanizing the connection between natural resources and provincial autonomy:

> From Alberta's vantage point, Ottawa's overall approach to the global energy price shock was viewed as an effective confiscation of their rents: revenues that ought to be flowing into provincial treasuries were being effectively transferred to Canadians in the form of subsidized domestic energy prices on the one hand and transferred to Ottawa

---

[8] Provincial governments could also argue that sections 92.10, "Local Works and Undertakings," 92.13, "Property and Civil Rights," and 92.16, "Matters of a merely local or private Nature" give them claim over the ownership, exploitation, and development of natural resources.

via the export tax on the other. And these forgone royalties soared as the difference between the world price and the domestic price likewise soared. (Courchene 2007, 26)

Although one could argue that the collapse of global oil prices in the early 1980s was more harmful to the province's economy than the actual provisions of NEP, this controversial federal policy still serves as a reference for those defending Alberta's resources, and autonomy, against federal actions viewed as detrimental. Indeed, "the NEP remains indelibly etched in the psyche of Albertans, ready to emerge when their interests are at stake" (Courchene 2007, 26).

Natural resources also play a direct role in the definition of the political community in Saskatchewan. Previously dominated by agriculture, it now self-identifies as an "energy and mineral powerhouse,"[9] enjoying its status as the world's largest producer of potash and second-largest producer of uranium in addition to being Canada's second-largest oil producer and third most important producer of natural gas. The attempted takeover of Potash Corp by BHP Billiton in 2010 demonstrated just how resources are meshed with Saskatchewan's identity, as popular premier Brad Wall was able to get the federal Harper government to nix the deal "by stoking an emotional response to an issue that was largely about provincial revenues" (Vanderklippe 2010). Potash has long been a staple of Saskatchewan's economy, yet the province's improved fortunes over the last several years are largely the result of expanded oil production in the 1990s combined with soaring oil prices. As such, even if Saskatchewan's resource sector remains more diverse (and recession-proof) than Alberta's, oil is becoming increasingly significant not only in the province's economy and fiscal policy, but also in its politics.

The same can be said about Newfoundland and Labrador and, to a lesser extent, Nova Scotia. In Newfoundland and Labrador, the development of offshore oil sites (Hibernia, Terra Nova, and White Rose) has generated tremendous revenues for the province and become symbolic of its transition from an equalization recipient to a non-recipient. These oil revenues are viewed as giving Newfoundland and Labrador a new status within the federation as an increasingly self-reliant province. Oil revenues from the Sable project offshore Nova Scotia have brought new wealth to that province and, in that context, forged links between resources and the political community.

In the case of Newfoundland and Labrador, hydroelectricity is also connected to provincial identity, as the Churchill Falls station meets much of the province's energy needs, with most of the leftover power sold to Hydro-Québec. The long-standing and politically charged struggle between the Newfoundland and Quebec governments over the transmission of hydroelectrical power produced in Labrador

---

[9] "Saskatchewan: Energy. Innovation. Opportunity," Fact sheet, n.d., http://saskfirstnations resources.ca/resource_development/oil_and_gas.html.

has contributed to connecting resources to provincial identity in Newfoundland. Indeed, Newfoundland and Labrador premiers have condemned the 1969 agreement with Quebec that earmarks most of Churchill Falls' power to Quebec, and in 2010 Danny Williams vigorously denounced a *Régie de l'Énergie du Québec* decision to deny transmission through Quebec of power that would come from the development of another hydroelectricity project in the Lower Churchill Falls. In this struggle with Quebec, the Newfoundland identity was a key element in a political fight over the control, exploitation, and transmission of renewable resources in the context of residual tensions over the physical boundaries of the two provinces (Churchill 1999). In March 2013, the Harper government put forward a $6.3 billion federal loan guarantee for the Muskrat Falls dam project, which should help hydro power remain a key source of economic development in Newfoundland and Labrador (CBC News 2012).

Hydroelectrical power became an important symbol of Québécois nationhood during the Quiet Revolution. In the early 1960s, as part of the recently elected Lesage government, natural resource minister René Lévesque championed the nationalization of hydroelectrical power. In 1962, the Quebec government went ahead with the purchase of all the private electricity companies doing business in the province, and Hydro-Québec, created in 1944, became the sole provider of electrical power in the province. The *nationalisation de l'électricité* quickly became a political symbol (the Quiet Revolution slogan was *Maîtres chez nous* [Masters of our own house]) of the type of state-driven transformative action the Quebec government was ready to undertake in the interests of French-speaking Quebecers. Hydroelectricity has proven an enduring reminder for Quebecers of their transition toward modernity, in addition to generating wealth and providing access to electrical power at below market prices. Hydroelectricity has also become a significant policy issue in provinces such as British Columbia and, especially, Manitoba, where a relationship between provincial identity and hydro power has emerged, notably through the development of the Crown corporation Manitoba Hydro.

Natural resources, renewable or not, are therefore extremely meaningful for provinces that have them in significant quantities. In particular, resources are a major source of revenue for these provinces. This is in part why provincial governments stress their ownership of natural resources whenever there is a political discussion around them. Moreover, resources have become in many provinces both a symbol and a concrete expression of a province's autonomy, individuality, and identity in a way that is coherent with the decentralist tendencies of Canadian federalism. Yet, as suggested above, there are also some territorial solidarity components to federalism in Canada, perhaps most notably equalization, and these often clash with the idea of provincial control and autonomy put forth by provincial governments when it comes to natural resources.

TERRITORIAL SOLIDARITY AND PROVINCIAL AUTONOMY: EQUALIZATION MEETS NATURAL RESOURCES

The tensions between equalization as a form of territorial redistribution and natural resources whose control lie with the provinces have come through in discussions and debates over the design of the equalization formula. As the 2006 report of the Expert Panel on Equalization and Territorial Financing Formula, commissioned in March 2005 by the Martin government to assess the equalization formula, asserted, "The treatment of resource revenues is the most complex and controversial aspect of Equalization" (5). This controversy is reflected in the many different treatments of non-renewable natural resources that have been featured in the various equalization formulas adopted since 1957.

Initially, the program excluded revenue from natural resources. In 1962, the federal government made 50 percent of revenue from non-renewable natural resources count in the calculation of provincial fiscal capacity, and it moved from an equalization standard based on the average fiscal capacity of the two wealthiest provinces to a national, ten-province average (a decision reversed only two years later). As a result of these changes, Alberta and British Columbia became non-recipient provinces, along with Ontario. In 1964, "the regulations concerning natural resource revenues were changed such that any province with a per capita yield from that source above the national average would have its equalization payments reduced by 50 percent of that amount" (Locke and Hobson 2004, 14). In 1967, 100 percent of revenues from non-renewable resources were included into the calculation of provincial fiscal capacity, and the equalization standard was made to be the (ten-province) national average once again. More changes occurred in the 1970s: "The regulations were again changed in 1974 so that only one-third of provincial oil and gas resulting from the increase in oil prices above the 1973 level would be subject to equalization. The balance – revenues calculated at 1973 prices – were to be fully equalized" (ibid.). In 1977, the federal government went back to the 50 percent inclusion rule. In 1982, it reverted to the full inclusion of non-renewable resources, but as the equalization standard was changed to rely on a five-province standard that excluded both the richest and the four poorest provinces from the calculation of the average fiscal capacity, Alberta (and its natural resources) was taken out of the equation. In 2004, the Martin government announced a New Framework that established a fixed pool for equalization and increased the total amount of money to be distributed through the program. Then in 2007, the Harper government, following the recommendations of the 2006 Expert Panel, opted for the inclusion of 50 percent of non-renewable-resource revenues and the return to a ten-province standard.

Considering the above remarks, there are three basic ways to deal with non-renewable natural resources in the context of equalization: full inclusion, full exclusion, and 50 percent inclusion. The argument for full inclusion is the most straightforward. Indeed, those who argue for this policy alternative see no difference between revenue from non-renewable resources and revenue from other sources. In other words, a dollar is a dollar, wherever it comes from. In addition, supporters of full inclusion make the point that non-renewable natural resources, due to their uneven territorial distribution, are a central source of interprovincial inequalities. Indeed, as oil and gas prices rise on world markets, interprovincial (territorial) discrepancies increase. High oil prices accentuate the fiscal capacity gap between provinces that have large quantities of non-renewable natural resources and the rest of the country. Hence, a program whose purpose is to address territorial inequalities should take into consideration the main source of these inequalities (Expert Panel 2006, 62). Unsurprisingly, recipient provinces with no (or limited quantities of) oil and gas have been the main proponents of this approach to non-renewable natural resources in the context of equalization policy. In the absence of a fixed pool, they gain by having non-renewable-resource revenues fully included in the equalization formula because it makes for a larger overall pot by boosting the average fiscal capacity. Full inclusion would now mean that Ontario would consistently be a recipient province, leaving less money for the "traditional" recipient provinces. From a normative and political perspective, regardless whether this term is used or not, the argument for full inclusion is about solidarity: no province should be able to shelter revenue from the redistributive mechanisms of Canadian fiscal federalism.

The argument for the full exclusion of non-renewables from the equalization formula is multidimensional but it begins with the idea of provincial autonomy. More specifically, it stresses the constitutionally specified legislative control of natural resources by provincial governments (Boessenkool 2002). In other words, provinces with non-renewable resources are saying that these resources are theirs, that they are not for sharing, and that including revenues from natural resources in the calculation of their fiscal capacity would involve a transfer of this wealth to other provinces.[10] When it comes to resources, the tension between equalization and provincial autonomy is not only about money but about powerful provincial identities associated with the strong belief that each province should control its resources and directly benefit from them, even if this means fighting against Ottawa.

This argument for full exclusion is perhaps more forcefully made by provinces such as Saskatchewan and Newfoundland and Labrador, which have experienced a "resource boom" in last few years. These traditional equalization recipients perceived exclusion from equalization payments on account of their recent resource boom as highly punitive and unfair. Indeed, there is a sense that the equalization program punishes provinces for developing their resources when this development

---

[10] Provinces with oil and gas also argue that these resources should be excluded from fiscal capacity calculation because they are non-renewable, that is, not part of an endless revenue stream.

generates revenues that push their fiscal capacity above the equalization standard. In reaction to the federal government's 2007 equalization reform, which featured an enriched pool but with 50 percent inclusion of resource revenues (a move that seem to go against Stephen Harper's previous position, which was to exclude these revenues altogether) and a cap on equalization (designed to make sure that a province's actual fiscal capacity post-equalization would not be greater than that of a non-recipient province), then Saskatchewan premier Lorne Calvert threatened to sue the federal government. The 2007 equalization reform, Calvert argued, was potentially unconstitutional because it disregarded provincial ownership of natural resources and contravened the clause stating that equalization payments should be equitable and fair (Canwest News Service 2007).

No province makes an argument for 50 percent inclusion of non-renewable natural resources in the calculation of the average provincial fiscal capacity. Yet, this is a formula that has been used in the past and, indeed, the one that is currently in use. This approach is a pure political compromise; equalization is, after all, closely tied to the politics of Canadian federalism (Lecours and Béland 2010). As the report of the Royal Commission on the Economic Union and Development Prospects for Canada stated, "A portion of resource revenues – greater than zero but significantly less than 100% – must be included in Equalization. There is no magic figure" (quoted in Expert Panel 2006, 58). In this context, 50 percent is a reasonable figure. The Expert Panel (2006) came to the same conclusion: "A portion of resource revenues should be included because of the fact that resource revenues do contribute substantially to a province's fiscal capacity" (57), but not all of these revenues should be factored in because, among other things, "based on the principle of policy neutrality, the Equalization program should not provide incentives or disincentives for provinces to develop natural resources" (57). The panel members also stated, "Our best judgement indicates that a 50 percent inclusion rate combines the merits of the various arguments and provides the most reasonable results for all receiving provinces" (58).

The 50 percent inclusion rule has overall been a good political compromise. It is certainly a better *political* solution to the issue of how equalization should deal with natural resources than full inclusion (which could be seen as unfair because the provinces would have no ability to take off the infrastructure and policy expenses of generating resource royalties in the first place) or exclusion, since both of these options would meet with much more forceful opposition on the part of some provinces. Yet, at least two issues have contributed to making this compromise politically problematic, at least for some provinces.

The first are the bilateral agreements on offshore resources between Ottawa and the governments of Newfoundland and Labrador and Nova Scotia, respectively. The express purpose of these controversial agreements, which are known as the Atlantic Accords, is to shield offshore oil and gas exploitation from equalization. In the case of Newfoundland and Labrador, the original 1985 Atlantic Accord stipulated that the government of the province could tax these resources as if they were its own (that is, as if they were on land), and that the federal government would make "offset payments" to Newfoundland and Labrador to compensate for a reduction in equalization resulting from the exploitation of the offshore resources. The renewed 2005 agreement between Ottawa and Newfoundland and Labrador looked to achieve a similar objective. It stated that "the Government of Canada intends to provide additional offset payments to the province in respect of offshore-related Equalization reductions, effectively allowing it to retain the benefit of 100 per cent of its offshore resource revenues." Similar agreements were struck between the federal government and the government of Nova Scotia in 1982 (Canada-Nova Scotia Agreement on Offshore Oil and Gas Resource Management and Revenue Sharing) and in 2005 (Offshore Revenues Agreement). In the two provinces, the argument has been that these accords are about economic development and have nothing to do with equalization policy; elsewhere in Canada, such "side deals" were viewed as involving a movement away from a formula-based equalization program (Expert Panel 2006, 28).

When the 2007 equalization reform was announced, the governments of Newfoundland and Labrador and of Nova Scotia were the most incensed by it as they saw the changes announced by Ottawa as breaking, at least in spirit, their bilateral accords with the federal government. The 50 percent inclusion of resource revenues into the calculation of provincial capacity combined with a cap that would stop payment when the overall fiscal capacity of a recipient province would reach that of a non-recipient province meant a substantial financial shortfall for the two provinces.[11] The politics of the 2007 reform were nasty in both provinces, especially in Newfoundland and Labrador, where Premier Danny Williams fiercely attacked the federal government. Even if, by 2008, the government of Newfoundland was proud to announce that it would soon no longer be a recipient of equalization payments, it still resented the changes made to the program by the Harper government. From the Newfoundland perspective, the federal government was penalizing the province for its success in developing its natural resources. The (offshore) "resource boom" in Atlantic Canada therefore represents as much of a political challenge to the federal equalization program as the Western "resource boom." In the provinces that have large quantities of non-renewable natural resources, any sense that the wealth stemming from their development leaves the provincial boundaries through the "visible hand" of Ottawa's equalization program triggers potential resentment and claims of unfair treatment.

The second issue that confronts the compromise of 50 percent inclusion of resource revenue is hydroelectricity. Equalization does not capture all revenues stemming from hydroelectricity because provinces such as Manitoba and Quebec

---

[11] The two provinces did have the choice to exclude all of their resource revenues from the calculation of their fiscal capacity if they chose to remain with the old, less generous program, which they opted not to do.

sell electricity to their residents at artificially low prices. As a result, the equalization payments of Manitoba and Quebec are boosted by the fact that some hydro revenues "disappear" through these below-market sale rates. As the *Globe and Mail* puts it in the case of Quebec, "The Quebec government artificially reduces its 'fiscal capacity' – thereby qualifying for higher equalization payments – by allowing provincially owned Hydro-Québec to charge consumers, especially large industrial ones, a price far below the market value" (Yakabuski 2008, B2). This is something that oil- and gas-producing provinces cannot do with their own resources. The treatment of hydroelectricity in equalization is seen as unfair by provinces that have large stocks of non-renewable resources. Saskatchewan, for example, has compared itself to its hydro-producing neighbour, Manitoba, and denounced the inclusion of its non-renewable resources into equalization and the exclusion of some of Manitoba's hydro resources (Couture 2012).

There are technical problems when it comes to the inclusion of hydroelectricity revenues into the calculation of provincial fiscal capacity (most notably that hydro power does not have international pricing like crude oil does), but these are likely not unsolvable. The biggest problems for provinces that would like to see these resources included is that such inclusion would involve difficult politics in Manitoba and, crucially, in Quebec. Hydroelectricity is viewed as a key force for economic development in both provinces and as a resource that is their exclusive property and whose benefits should be enjoyed solely by the province's residents. Of course, this is virtually the same discourse as the one coming from oil- and gas-producing provinces about their own (non-renewable) resources. Yet, as we have seen, the status of oil and gas within equalization has changed over the years, which means that 50 percent inclusion is somewhat palatable. In the case of hydroelectricity, Manitoba and Quebec would point out that moving to take into account revenues "hidden" in the discounted sale of power to residents lacks precedent. In Quebec, where hydro power is closely linked with ideas of nationhood and autonomy, such a policy change would be particularly controversial.

CONCLUSION

There is a basic political and ideological tension between equalization and natural resources in Canada: equalization is a concrete expression of territorial solidarity, whereas natural resources are strongly associated with the idea of provincial autonomy. Provincial governments typically do not see natural resources in a pan-Canadian perspective; rather, they consider that resources are not for sharing, as they need to benefit their own residents. Therefore, when it comes to equalization and resources, resource-rich provinces do not "buy in" to a logic of redistribution (Hirsch 2012). In this political context, there has been little interprovincial cooperation when it comes to natural resources in the country.

Yet, interestingly, the recent Northern Gateway project has been at odds with this pattern, as former Alberta premier Allison Redford promoted the notion of a National Energy Strategy. The very wording of this initiative may sound odd as something coming from Alberta, not only because it can recall the arch-demonized NEP but also because provincial governments, as we have seen, tend to avoid placing resources in a pan-Canadian framework. However, for the current Alberta government, the challenge is to get oil over to Asia. This involves some form of cooperation with the British Columbia government, which responded to the Northern Gateway project by asking for a "fair share" of royalties while voicing environmental concerns. In more general terms, the Northern Gateway has meant an otherwise surprising effort at "Canadianizing" energy policy in Canada. Premier Redford brought her idea of a National Energy Strategy to the 2012 Council of the Federation (COF) meeting in Halifax and spearheaded the renewal of the 2007 COF initiative on energy, a process from which the BC government disengaged because of Alberta's refusal to compensate it in any way for pipelines going through its territory. Indeed, Liberal BC premier Christy Clark indicated that she was not interested in discussing any national energy strategy.

The Alberta government is therefore making appeals to Canadian solidarity so that its oil can be shipped to Asia. It has emphasized interprovincial cooperation because it is, at this point in time, in the province's interests. Interestingly, this appeal has been well received by most provinces, especially land-locked ones such as Saskatchewan, which see some potential threats to their export capacity in the position of the BC Clark government. Back in Alberta, Wildrose leader Danielle Smith was, as expected, critical of Premier Redford's plan, linking it to the NEP and arguing that Canada's energy successes have occurred without a "pan-provincial, multi-jurisdictional, comprehensive Canadian Energy Strategy" (Smith 2012). The federal government was, for its part, strangely mute about an initiative labelled as "national."

While the notion of a Canadian Energy Strategy emphasizes interprovincial cooperation, the connection between natural resources and provincial autonomy remains paramount. For example, no provincial government supported British Columbia in its claim that it should receive a "fair share" of royalties for oil pipelines going through its territory on the way to Asia from Alberta. The conventional thinking about resources in Canada is that a province's resources should benefit its residents, almost to the exclusion of all other Canadians, a notion that hurts the logic of territorial redistribution at the core of the federal equalization program.

REFERENCES

Boessenkool, Kenneth J. 2002. *Ten Reasons to Remove Nonrenewable Resources from Equalization*. Halifax: Atlantic Institute for Market Studies, Frontier Centre for Public Policy, Montreal Economic Institute.

Bryden, P. E. 2009. "The Obligations of Federalism: Ontario and the Origins of Equalization." In *Framing Canadian Federalism: Historical Essays in Honour of John T. Saywell*, edited by Dimitry Anastakis and P.E. Bryden, 75-94. Toronto: University of Toronto Press.

Burgess, Michael, and John Pinder. 2007. *Multinational Federations*. London: Routledge.

Canwest News Service. 2007. "Saskatchewan to Sue Federal Government over Resource Revenues." *Canada.com*, June 13. http://www.canada.com/calgaryherald/news/story. html?id=36339e71-a22a-433b-b43a-6487ab21175d&k=78694.

CBC News. 2007. "N.L. 'Shafted' in Federal Budget: Williams." *CBC News.ca*, March 20. http://www.cbc.ca/news/canada/newfoundland-labrador/story/2007/03/20/william-budget-reax.html.

—. 2012. "Muskrat Falls Hydroelectric Project Clears Major Hurdle: Stephen Harper Announces $6.3B Loan Guarantee in Labrador Friday." *CBC News.ca*, November 30. http://www.cbc.ca/news/canada/newfoundland-labrador/story/2012/11/30/nl-muskrat-falls-loan-guarantee-1130.html.

Churchill, Jason L. 1999. "Pragmatic Federalism: The Politics behind the 1969 Churchill Falls Contract." *Newfoundland and Labrador Studies* 15 (2): 215-46.

Courchene, Thomas J. 1984. *Equalization Payments: Past, Present and Future*. Toronto: Ontario Economic Council.

—. 2007. "A Short History of Equalization." *Policy Options* (March): 22-29.

—. 2012. *Policy Signposts in Postwar Canada: Reflections of a Market Populist*. Montreal: IRPP.

Couture, Joe. 2012. "Wall Welcomes Possible Changes to Equalization Program." *Star Phoenix*, October 11.

Expert Panel on Equalization and Territorial Financing Formula. 2006. *Achieving a National Purpose: Putting Equalization Back on Track*. Ottawa.

Hirsch, Todd. 2012. "We Must Buy Into Regional Redistribution." *Globe and Mail*, October 22, A13.

Janigan, Mary. 2012. *Let the Eastern Bastards Freeze in the Dark: The West Versus the Rest since Confederation*. Toronto: Knopf Canada.

Leblanc, Daniel. 2012. "CAQ Leader François Legault Puts Economy First." *Globe and Mail*, August 27.

Lecours, André, and Daniel Béland. 2010. "Federalism and Fiscal Policy: The Politics of Equalization in Canada." *Publius: The Journal of Federalism* 40 (4): 569-96.

Locke, Wade, and Paul Hobson. 2004. *An Examination of the Interaction between Natural Resource Revenues and Equalization Payments: Lessons for Atlantic Canada*. Montreal: IRPP.

Milne, David. 1998. "Equalization and the Politics of Restraint." In *Equalization: Its Contribution to Canada's Fiscal and Economic Progress*, edited by Robin W. Boadway and Paul A.R. Hobson, 175-203. Kingston: John Deutsch Institute for the Study of Economic Policy.

Pierson, Paul. 1996. "The New Politics of the Welfare State." *World Politics* 48 (2): 143-79.

Royal Commission on Dominion-Provincial Relations. 1940. *Report of the Royal Commission on Dominion-Provincial Relations. Book II: Recommendations*. Ottawa: King's Printer.

Séguin, Rhéal. 2004. "Ottawa's Policies Put Federalism at Risk: Charest." *Globe and Mail*, November 15, A7.

Smith, Danielle. 2012. "Why We Don't Need a Canadian Energy Strategy." Speech to the Economic Club of Canada, Ottawa, March 8. http://www.wildrose.ca/speech_to_the_economic_club_of_canada.

Thompson, Andrew R. n.d. "Resource Rights." *The Canadian Encyclopedia*. http://www.thecanadianencyclopedia.com/articles/resource-rights.

Tupper, Allan, Larry Pratt, and Ian Urquhart. 1992. "The Role of Government." In *Government and Politics in Alberta*, edited by Allan Tupper and Roger Gibbons, 31-66. Edmonton: University of Alberta Press.

Vanderklippe, Nathan. 2010. "How Brad Wall Turned Public Opinion against the Potash Deal." *Globe and Mail*, November 4. http://www.theglobeandmail.com/news/politics/how-brad-wall-turned-public-opinion-against-the-potash-deal/article1241327/.

Yakabuski, Konrad, 2008. "Have-Not Status Is All about Gaming the Rules." *Globe and Mail*, November 6, B2.

IV

Resources and Resiliency: Institutional Considerations

7

COMPARING INTERGOVERNMENTAL INSTITUTIONS IN HUMAN CAPITAL DEVELOPMENT

Donna E. Wood

INTRODUCTION

As outlined by many of the contributors to this 2012 State of the Federation volume, regional tensions in Canada remain alive and thriving. Problems related to human capital development – an aging workforce, skills and labour shortages, unemployment and underemployment, qualifications recognition and transferability of credentials across Canada and internationally, income inequality, poverty, access to post-secondary education, exclusion of citizens from the labour force (including Aboriginal people, people with disabilities, and immigrants), work-life balance, access to child care, and the need to enhance the competitiveness of the Canadian workforce – are felt differently in different places across the country. And there is often disagreement as to whether the issue should be tackled at the provincial/territorial level or on a pan-Canadian basis.

Historically we have managed our federal-provincial and interprovincial tensions by using a variety of intergovernmental relations (IGR) institutions. These include multilateral federal/provincial/territorial and provincial/territorial first ministers, ministers, deputy ministers, and officials' *forums* that provide an opportunity for dialogue, debate, and decision-making on a pan-Canadian basis. Intergovernmental *agreements* (both bilateral and multilateral) often detail the outcome of this dialogue

This chapter is based on my experience working in intergovernmental relations in human resources for Alberta between 1993 and 2003, interviews and correspondence with those directly involved in 2012–13, and a 2013 review of relevant websites.

by outlining more specific undertakings between governments. By collecting information, providing analysis, and undertaking research, *intermediary organizations* – whether directly funded by governments or supported by memberships, donations, or private foundations – are additional institutions that can act as a catalyst and connector to facilitate dialogue and understanding between governments, as well as with citizens from across the country.

However, over the past 20 years intergovernmental relations in Canada and the institutions that support it have fallen beneath the radar screen. Where once first ministers met regularly to discuss constitutional reform, the social union, and the state of the Canadian economy, since 1992 these conversations have taken place mostly on an informal or bilateral basis. The one constant has been regular meetings of premiers, recast in 2003 as the Council of the Federation. The number of formal federal/provincial/territorial meetings supported by the Canadian Intergovernmental Conference Secretariat has declined overall, from an average of more than 100 per year pre-2006 to an average of 70 per year since (CICS 2013).[1] As a result, many federal-provincial deliberations are being handled by senior officials, out of sight and hidden from public scrutiny.

In human capital matters, the work has traditionally been divided by policy sector into four groups: (1) Ministers of Labour and the Canadian Association of Administrators of Labour Legislation (CAALL), which deals with labour relations and workplace health and safety; (2) the Council of Ministers of Education Canada (CMEC), which deals with primary, secondary and post-secondary education; (3) the Forum of Labour Market Ministers (FLMM), which deals with workforce development, employment, and labour mobility; and (4) Ministers of Social Services (MSS), which deals with social assistance, poverty, disability, and welfare services.[2]

The purpose of this chapter is to compare the institutionalization of intergovernmental relations in these different policy domains between 1993 and 2005 (under Liberal control), and then under the Harper Conservatives between 2006 and 2012. I also examine policy capacity, that is, the ability of Canadian governments to work together to address public policy issues. To put Canada's IGR practices in a comparative context, I bring in information on how employment policy is coordinated across the European Union's (EU) 28 member states. The EU is considered as a legitimate comparator to Canada because intergovernmental federalism is the operative mechanism in both political systems (Hueglin 2013). The relevant EU intergovernmental forum is called the Employment, Social Policy, Health and Consumer Affairs Council (EPSCO).

---

[1] See in particular *Reports to Governments* 2011–2012, 2010–2011, 2009–2010, 2008–2009, and 2007–2008 and *Strategic Plan* 2013–2018.

[2] The immigration sector is a fifth federal/provincial/territorial forum in human capital matters. It is not reviewed in this chapter.

I start by considering some of the key characteristics of intergovernmental relations in Canada, and then outline the framework I will use to measure and compare IGR institutionalization and policy capacity. This is followed by a brief overview of how the four Canadian forums and associated institutions actually function. I then compare the policy domains using the following elements from the analytical framework: interdependence in the sector, nature and composition of the forums, secretariat and functioning, relationship to first ministers or heads of state, presence or absence of intergovernmental agreements, transparency, participation beyond government executives and involvement of intermediary organizations, and outcomes/outputs in the sector. EU approaches are then considered using the same framework as in the Canadian assessment.

The chapter concludes with a summary table comparing the five intergovernmental forums, and an analysis of whether the structures in place in Canada succeed in facilitating constructive dialogue on national policy issues relating to the development of Canada's human resources. The chapter argues that the short answer to this question is no. While particular aspects of the Canadian labour market and current trends cry out for intergovernmental responses, our IGR machinery is generally inadequate to the task and is getting increasingly weaker, especially since the Harper Conservatives assumed control in 2006. While there are many improvements that could be considered, public interest and political will would be required to engage with the problem in order to move any solutions forward.

HUMAN CAPITAL DEVELOPMENT RESPONSIBILITIES IN CANADA

In Canada, provinces score high for self-rule (regional government authority over those who live in their territory) but lower for shared rule (capacity of regional governments to shape national decision-making; Hooghe, Marks, and Schakel 2010). This results in built-in challenges to pan-Canadian coordination, what Steven Kennett (1998) has called our "collective action problem."

The 1867 British North America Act assigned most matters relating to human capital development to the provinces. However, after the Second World War, with the agreement of all provinces (and sometimes at the urging of the federal New Democrats), the governing federal Liberals were successful in overcoming our collective action problem in order to build the Canadian welfare state. This included constitutional amendments expanding the competence of the Government of Canada by transferring responsibility for unemployment insurance and pensions to the federal level. Ottawa also helped provinces expand social programs under their constitutional authority (health care, post-secondary education, social assistance, and social services) using the federal spending power.

While provinces still retained most social policy responsibilities, these actions greatly increased federal-provincial interdependence in human capital development beyond what was outlined in our founding constitutional documents in 1867 (Banting 2005). This interdependence was not substantially changed under various Progressive Conservative governments that ruled at different times between 1957 and 1993. However, since assuming power in 2006, the Harper Conservatives under a policy called "open federalism" – intended to limit the federal spending power, clarify roles and responsibilities, and respect areas of provincial jurisdiction – have taken a different approach. This political positioning has had a profound impact on Canada's intergovernmental institutions and their operation.

CONCEPTUALIZING AND MEASURING INTERGOVERNMENTAL INSTITUTIONALIZATION AND POLICY CAPACITY

Constituent units in any federation are bound together by common rights and obligations, as well as by bonds that develop among actors and communities. Even though our constitutional bargain divided authority by subject area between orders of government, there are many issues where entanglement, overlap, and interdependence are inevitable because of funding, administration, or the "transversal" nature of the policy area.

Why Governments Interact

Irrespective of the times or the particular philosophy of the government in power, governments in a federation need to interact for a variety of reasons: to shape a policy direction together either strategically or operationally in order to solve an agreed-upon problem, because neither government has the necessary constitutional powers and/or financial resources; to implement or uphold a policy direction and account for action; to share resources to act in a mutually beneficial way to achieve common goals; to coordinate action to ensure overall coherence and harmonization; to clarify roles and responsibilities and reduce overlap and duplication; and to undertake research, compare what works, and exchange information for policy learning. Many of these intergovernmental objectives are motivated by a strong concern for the effective delivery of public services to citizens and are worked through/resolved at the sector level, not through "high politics" or the media.

But intergovernmental relations are also needed to prevent surprises from unilateral action or to adjust actions, to influence behaviour or persuade a party to act in a certain way, to challenge the behaviour and action of a party, to prevent certain actions and subsequent negative consequences, to protect or advance jurisdiction, and to resolve conflicts and disputes. Since in Canada we lack a legitimate and effective Senate as a way for provinces to participate in legislative decision-making at the centre, it is our intergovernmental relations system that does the heavy lifting in this regard. Without a meaningful intergovernmental relations system that respects self-rule and power sharing, any federal political system would tip over into a unitary or confederal system (Agranoff 2004).

Types of Intergovernmental Relations

The most basic form of intergovernmental relations is *voluntary mutual adjustment* or *ad hoc coordination* through informal means. This does not require regular meetings, a bureaucratic structure, decision-making rules, or formal agreements, and it allows for maximum flexibility and autonomy of the participating partners. *Bilateralism* is by far the preferred approach to managing intergovernmental relations in Canada; Poirier (2001) estimates that 85 percent of the 1,500–2,000 existing intergovernmental agreements are bilateral. Such arrangements range from a unique agreement between the federal government and an individual province, to similar agreements between Ottawa and each province, to agreements between two provinces.

Multilateralism involves governments acting or trying to act collaboratively: Ottawa and all of the provinces and territories together (federal/provincial/territorial relations); all of the provinces and territories without the federal government (inter-provincial/territorial relations); or just some of the provinces and territories (regional relations). In practice, many multilateral arrangements are between Ottawa and all provinces/territories except for Quebec, which often views multilateral agreement as interference in its jurisdiction. Multilateral cooperation among provinces and territories is often difficult to achieve as each has its unique objectives and interests. The requirement for unanimity and joint agreement frequently detracts from overall policy effectiveness. On the other hand, multilateralism allows provinces and territories to present a common front vis-á-vis the federal government, enables better coordination across Canada, and provides a mechanism for the consideration of pan-Canadian approaches.

Measuring Intergovernmental Institutionalization

Formal governance rules enable cooperation because they create stable expectations about behaviour, and signify common values for those involved in an exchange. Such rules also enhance the credibility of commitments, making people more willing to cooperate. Bolleyer (2006) compared the institutionalization of IGR in six federations – Germany, Switzerland, United States, Austria, Australia, and Canada – by looking at peak institutions (e.g., First Ministers and the Council of the Federation). She concluded that Canada had very low scores related to institutionalization. However, peak institutions are just the tip of Canada's IGR system.

Drawing on Dennison (2005) and Bolleyer (2006), this chapter uses the following variables to assess the institutionalization of the four policy sectors in Canada and one of these sectors in the EU: a founding agreement, statement of purpose, or mandate; presence or absence of senior level (minister or deputy minister) engagement; an established pattern of meetings; shared or rotating chairmanship; secretariat support with defined funding arrangements and defined functions; subgroup working structure; presence (or absence) of defined intergovernmental agreements; degree of transparency, including the presence of a website and information on activities and outcomes; degree of participation beyond government executives; presence of formal decision rules that deviate from consensus; and presence of intermediary organizations (for research, information, analysis, and for facilitating dialogue, knowledge exchange, and information dissemination).

Measuring Intergovernmental Policy Capacity

The degree to which a sector is institutionalized does not help us understand its policy capacity. Dupré (1985) defined a *workable* IGR system as one that nurtures institutions and processes "conducive to negotiation, consultation, or simply an exchange of information" (233). Bakvis and Skogstad (2012) used workability to assess the "performance" of Canada's IGR system, that is, "the capacity of federal institutions to produce results in the form of agreements" (3). When in 2011 Inwood, Johns, and O'Reilly assessed intergovernmental relations in finance, trade, health, and environment policy in Canada, they argued that researchers needed to go a step further and examine policy capacity, "the ability of federal and provincial governments in Canada to work together to address public policy problems" (14). Policy end results and policy output are important, not just process. It is with this understanding that I attempt to assess policy capacity in this chapter.

OVERVIEW OF CANADIAN INTERGOVERNMENTAL INSTITUTIONS AND PROCESSES BY SECTOR

As already outlined, there are four key human capital intergovernmental institutions in Canada. A brief overview of each is presented below.

Ministers of Labour and the Canadian Association of Administrators of Labour Legislation (CAALL)

CAALL is the longest-running intergovernmental institution of those examined, and the most formalized. Established as a federal-provincial body in 1938, with its formal constitution last amended in 2011, the association recently celebrated its 76th annual meeting. The deputy ministers of labour normally meet twice a year, and the ministers meet annually. This practice has not changed since the Conservatives assumed power in 2006. Since only 10 percent of the Canadian workforce is under federal jurisdiction, Canada has 14 different sets of labour laws and 14 independent administrative structures for industrial relations, employment standards, occupational safety and health, and workers' compensation.

The main objectives of CAALL are to encourage cooperation among members, provide a forum for exchange of experiences, encourage research, and promote high standards of administration. CAALL interacts with the International Labour Organization, and also works on international health and safety issues. It coordinates Canadian activities with the National Association of Governmental Labour Officials, its American counterpart. The CAALL executive (at the deputy minister level) is composed of the president, four vice-presidents, the secretary, and the past president. The provincial president rotates. A permanent secretariat with one full-time-staff equivalent is provided by and funded through the labour program of Human Resources and Skills Development Canada or HRSDC (renamed Employment and Social Development Canada or ESDC in 2013). There are five standing committees at the level of officials.

Council of Ministers of Education Canada (CMEC)

CMEC is unique among all the intergovernmental forums examined in this chapter as it is exclusively a provincial/territorial body. Covering early childhood learning and development, elementary and secondary schooling, post-secondary education, and adult learning and skills development, CMEC serves as a forum to discuss policy issues and undertake activities in areas of mutual interest. The Council consults and cooperates with national education organizations and the federal government, and represents the education interests of the provinces and territories internationally. CMEC's international component includes implementing Canada's obligations under UNESCO conventions, and working with the OECD to compare Canada's education system to that in other developed countries. CMEC was formed in 1967 to assert provincial jurisdiction and competence over education matters. Provinces – Quebec and Alberta in particular – patrol the boundaries and guard provincial jurisdiction on an ongoing basis; the federal government is invited to engage only on very specific matters, mostly on an ad hoc basis. Haskel (2013) calls federal involvement in post-secondary education the "elephant in the room," laying down puzzle pieces (e.g., student assistance, research, international marketing) without consultation with the provinces. Without routinized and regular access to CMEC, the federal government has undertaken unilateral action on many post-secondary education issues.

CMEC is highly institutionalized, with work driven at the level of ministers and deputy ministers, who hold face-to-face meetings two or three times per year. In July 2014, ministers celebrated their 103rd CMEC meeting. Located in Toronto, the secretariat is headed by a director general supported by 60 staff. CMEC is governed by an Agreed Memorandum approved by all members. It is funded 75 percent by levies on provincial governments and 25 percent by the Government of Canada. It has an executive of five provinces, with a chair elected every two years based on rotation among the provinces. The CMEC secretariat contains the Canadian Education Statistics Council, a partnership between CMEC and Statistics Canada, and the Canadian Information Centre for International Credentials, which facilitates the recognition of foreign credentials.

Forum of Labour Market Ministers (FLMM)

The FLMM was set up in 1983 as a federal/provincial/territorial forum to promote interjurisdictional cooperation and establish common goals on labour market issues, to promote a highly skilled portable workforce, to facilitate adaptation to changes in skill requirements, and to provide a link to labour force development boards. These boards have since been disbanded and the FLMM objectives have changed over time. The forum's current work is focused on the mobility of workers (including foreign qualifications recognition), effective employment services, workforce development, and labour market information. Its international dimension is relatively minor.

The forum is co-chaired by ESDC and a lead province; this rotates every two years. An equal split of federal and provincial/territorial resources provides funding for the lead province to hire two or three staff to perform a secretariat role. Most work is done by assistant deputy ministers (called FLMM senior officials) via teleconferences held six to eight times per year. There are five subofficials' working groups. Deputy ministers generally meet annually. Ministers met rarely in the past decade, only in 2010 and 2003. Since the Conservatives assumed power in 2006, there has generally been reduced activity at the deputy and ministers' levels. However, in 2013 the federal government decided to unilaterally alter the parameters of a key federal-provincial agreement in order to create a Canada Job Grant. This reactivated the forum, especially at the provincial level. The Canadian Council of Directors of Apprenticeship (CCDA) also plays a major role as an intergovernmental body in labour market matters, and facilitates research through the Canadian Apprenticeship Forum.[3]

Intergovernmental relations in labour market matters are mainly structured through four different types of labour market agreements that, starting in 1996, devolved federal funding to provinces and territories for active measures.[4] Negotiated bilaterally, there are 49 different agreements, representing an annual transfer from Ottawa to the provinces of approximately $2.8 billion. Despite devolution, Ottawa continues to be directly responsible for managing active measures for Aboriginal persons, youth, and people with disabilities.

Ministers of Social Services (MSS)

This forum started in 1975 as a provincial/territorial activity; the federal government became involved in 1985. It has a very limited international dimension. Federal funding cutbacks to provincial social programming in 1996 created considerable turmoil in the intergovernmental environment. When financial stability was re-established by the late 1990s, the Liberals were anxious to reinvest in social policy matters, both directly and through provincial transfers. In response the forum became very active, developing the National Child Benefit and In Unison (for persons with disabilities), as well as a National Children's Agenda. MSS also provided significant input to the Social Union Framework Agreement (SUFA) negotiations, which established a "code of conduct" for intergovernmental relations in social policy matters. Throughout this period ministers and deputy ministers met face-to-face twice a year, supported by officials' working groups and a director-level Support Committee.

The work of this forum substantially changed after the Conservatives came to power in 2006. As they view provinces as responsible for social policy, federal engagement through ESDC is limited. There have been no federal/provincial/territorial ministers' meetings since the cancellation of the early learning and child-care agreements in 2006. Deputy ministers meet via teleconference or face-to-face about

---

[3] CCDA has been in place since 1958 to facilitate the mobility of tradespersons across Canada through the interprovincial "Red Seal" program. At one time it reported to FLMM deputy ministers. CCDA is very active, with provincial/territorial directors meeting twice a year. The secretariat is supported by ESDC. The Canadian Apprenticeship Forum, established in 2000, is a national body that brings together all players in apprenticeship training: http://www.caf-fca.org/index.php. It is funded by the ESDC sector council program.

[4] Labour Market Development Agreement, Labour Market Agreement, Targeted Initiative for Older Workers, and Labour Market Agreement for Persons with Disabilities. The latter represents an updating of a previous federal-provincial agreement.

once a year. The federal/provincial/territorial Support Committee meets mostly by teleconference, with three defined working groups in operation. The lead-province role rotates every two years; provinces collectively contribute the equivalent of one position for overall coordination. There are also director-level federal/provincial/ territorial committees on income support, child welfare, research, and child care that try to meet annually. The federal government provides funding to cover 50 percent of the cost of federal/provincial/territorial meetings.

COMPARING INTERGOVERNMENTAL INSTITUTIONS IN CANADA

I turn now to a direct comparison of the four Canadian IGR institutions and pro- cesses, focusing on key dimensions from the analytical framework: interdependence in the sector, nature and composition of the forums, secretariat and functioning, relationship of the work to first ministers or heads of state, presence or absence of intergovernmental agreements, degree of transparency, participation beyond government executives and involvement of intermediary organizations, and out- comes/outputs in the sector.

Interdependence in the Sector

Across the four policy domains, federal-provincial interdependence is greater in the labour market and post-secondary education sectors. In labour market matters, governments acknowledge their interdependence and agree that competence is, in effect, shared. Employment Insurance is a federal constitutional responsibility. Ottawa also provides a $2.8 billion annual contribution through conditional bilateral agreements to cover the costs of active labour market measures delivered by prov- inces and territories. Both orders of government offer parallel programming (for youth, Aboriginal persons, and persons with disabilities) that requires coordination in order to be effective. In education matters, the presence of a significant inter- national component, as well as the involvement of both governments in student assistance, makes interdependence a factor in this sector as well. While in K–12 matters federal-provincial interdependence is more limited (except in regard to Aboriginal education), interdependence looms large in post-secondary education where Ottawa directly funds and manages one portion of the system (research) while provinces fund and manage another (teaching).[5]

---

[5] Ottawa also provides resources to provinces for post-secondary education (along with social assistance and social services) through the Canada Social Transfer. Since the transfer is unconditional, interdependence is more limited.

There is less interdependence in the social services and labour sectors, where governments agree that most responsibilities are provincial. In labour matters, the most that has ever been aspired to under both Liberal and Conservative governments is a sharing of information and best practices. This contrasts with social services, where historically the federal Liberals saw the Government of Canada as playing a coordination, leadership, and funding role, even though most programs were in provincial jurisdiction. The Conservatives view the world differently, hence the cancellation of the early learning and child-care agreements, the hiatus on further federal investments in the National Child Benefit, and a lack of engagement on disability issues. Many social policy advocacy organizations, it should be noted, fundamentally disagree with this approach; in their view, issues such as income inequality require a pan-Canadian approach *with* the engagement of the federal government.[6]

Nature and Composition of the Forums

Three of the four forums contain federal representatives as well as provincial/ territorial governments. The exception here is CMEC; whether under Liberal or Conservative federal leadership, provinces and territories in Canada have consistently resisted institutionalizing how they connect with the federal government under the rationale that education is solely within provincial jurisdiction (Cameron 2005). This position may indeed find some degree of accuracy as it relates to K–12 education, but it is not the case for post-secondary education, where the federal government has played a long-standing role. Even though CMEC (2003) itself commissioned an internal report that recommended finding productive ways to enter into federal/provincial/territorial dialogue, and the Canadian Council on Learning (2011) made a detailed proposal to set up a Ministers' Council on Learning, no action has been taken. Haskel (2013) suggests that changing approaches to intergovernmental coordination in post-secondary education would require a compelling vision, objective, or anxiety in order to motivate the key actors.

The most formalized IGR institutions in Canada are the oldest: CAALL and CMEC. Both have constitutions or memorandums of understanding outlining how they operate, and ongoing routinized face-to-face engagement at both the deputy ministers' and the ministers' levels. This engagement has been in existence for many years and is now embedded in practice. To *not* have ministers' meetings would be an exception. Both policy domains also have a significant international

---

[6] See the 47 submissions to the Parliamentary Committee Study of Income Inequality April 2013, available at http://www.parl.gc.ca/HousePublications/Publication.aspx ?DocId=6079428&Language=e&Mode=1&Parl=41&Ses=1.

dimension, requiring a coordinated interprovincial response to defined issues. Both have well-developed websites and permanent secretariats.

Next most formalized and institutionalized is the FLMM, with active working groups and commitments from all jurisdictions to fund the forum's ongoing activities. The Social Services forum is the least institutionalized. What is noteworthy about these two forums (compared to CMEC and CAALL) is the level at which activity takes place. Until 2013 when the Conservatives proposed a Canada Job Grant, ministers in both forums almost never met, and deputy ministers' meetings were rare. Almost all activity took place among officials through working groups; as a result, it is difficult to know what is happening as information is not made publicly available.

All multilateral IGR work, irrespective of the sector, progresses through consensus. Only with respect to the Canada Pension Plan (a jointly managed federal-provincial program) are there different decision rules. If an individual province (usually Quebec) does not agree with the overall direction being taken, a footnote is usually placed in the communiqué to identify this dissent. One respondent noted that achieving and then holding consensus is one of the most challenging aspects of IGR in Canada, especially given the rapid turnover of provincial players as governments in power change at different times.

Secretariat and Functioning of the Forums

All of the forums have a provincial chair or co-chair that rotates from one jurisdiction to another based on a predefined pattern. This position provides oversight to the secretariat. CMEC clearly has the strongest secretariat. Many who work for CMEC perform the work of an "intermediary organization," hence its large size. The federal government provides the CAALL secretariat, ensuring stability and institutional memory. Neither secretariat decides what the forum is to do; its role is to carry out the wishes of the executive committee, dominated by provincial governments.

The rotating provincial FLMM and Social Services secretariats are the weakest model; however, the FLMM provides slightly more resources than the Social Services forum where one staff equivalent (as opposed to two or three) provides overall coordination. The problem with a rotating secretariat is that it is completely dependent on the competence, capacity, and interest of the provincial co-chair. This can especially be a problem for underresourced, smaller provinces. With a rotating secretariat institutional memory is lost, as well as capacity to carry over projects that take longer than 18 months. Both forums undertake very limited research and analysis, undermining policy capacity of the sector overall. Federal support to these two secretariats has diminished since the Conservatives came to power in 2006.

Relationship to First Ministers or Heads of State

The work that ministers of social services and the FLMM undertook to feed into the federal/provincial/territorial social union negotiations in the late 1990s was triggered by action at the first ministers' level. Labour ministers and CMEC were not as directly involved. This activity reached its peak through the establishment of the Ministerial Council on Social Policy Reform and Renewal in 1999 under the federal Liberals, but like the Social Union agreement itself, the work has since faded away.

Since 2006 and the assumption of power by the Conservatives, there have been few issues where any of the four forums has undertaken action in response to requests from federal/provincial/territorial first ministers. Premiers' discussions at the Council of the Federation meetings in 2006, 2009, 2011, and 2013 triggered FLMM and CMEC action on labour mobility, foreign qualification recognition, and postsecondary education and skills training. This included requests to the Government of Canada to engage. However, there has been no response. Prime Minister Harper has called only one formal First Ministers' Meeting since he took power in 2006, and refused the premiers' invitation in the fall of 2012 to attend their meeting to facilitate cooperation on a national economic strategy. Connections are handled instead on a bilateral and ad hoc fashion (Whittington 2012).

Intergovernmental Agreements

The FLMM is the outlier in this respect, as the policy sector functions, in effect, through the 49 bilateral federal/provincial/territorial funding agreements. In the past decade, none of the FLMM working groups appeared to be focusing on these agreements, despite calls to consolidate and simplify the arrangements (Wood and Klassen 2012). However, this seems to have changed in light of the 2013 and 2014 federal Conservative budget announcements signalling a desire for change to the Labour Market Agreement in order to implement the Canada Job Grant. Ottawa has also signalled a desire to change the parameters of the more far-reaching Labour Market Development Agreements. Premiers have consistently expressed concern that the proposed federal changes would jeopardize the success of the training programs they already have in place (Council of the Federation 2013).

CMEC's agreements with Statistics Canada with respect to the Canadian Education Statistics Council, and a protocol with the Government of Canada on official languages in education, structure relationships in these areas on an ongoing basis. Agreements that dominated the social services sector in the 1990s and 2000s under the federal Liberals are now mostly in the past. The National Child Benefit agreement was implemented through joint press releases. Quebec has a unique agreement with Ottawa regarding parental leave, allowing the province to use funds from the Employment Insurance account to fashion a leave program for parents on the birth of their child that meets Quebec's needs. No other province has taken up this arrangement.

Transparency

A very significant difference between the four forums is the amount of information available, especially through public websites. Because CMEC ministers meet regularly and there is a robust secretariat able to undertake research and analysis, considerable information is available on overall directions, actions, and outcomes. CMEC's public website contains extensive and up-to-date information on its activities and deliverables, including 230 publications and reports (CMEC 2013). Although CAALL has a public website maintained by the federal government (CAALL 2013), up-to-date deliverables and current activities are provided on the private, members-only side of the website. Working groups are listed, but there are few deliverables and the information is dated.

Despite a limited secretariat, the FLMM maintains two websites on labour market information and labour mobility, as well as a broader website for the forum as a whole (FLMM 2013). The latter was set up by Alberta in 2012, with the assistance of federal ESDC officials. Unfortunately, much of the information on the three sites is dated. Since the forum does no research or outreach, the information available to be posted is slim. There is no cross-referencing to the bilateral agreements and reports available on ESDC's site. There is no website or information publicly available at all on the work of the Social Services forum. Although the National Child Benefit (2013) website is still operational, the latest reports refer to 2008 activity. A federally run website that connected to the Social Union Framework Agreement, the National Child Benefit, In Unison, and the child-care agreements (all federal Liberal initiatives) quietly disappeared a few years ago, after the Conservatives assumed power in 2006.

Participation beyond Government Executives and Involvement of Intermediary Organizations

All of the IGR forums discussed in this chapter are limited to participation by government officials; although the CMEC secretariat may not be composed of government officials, it acts on behalf of provincial/territorial governments. However, from time to time the forums actively reach out to those outside of government. CAALL will on occasion invite representatives of the Canadian Centre for Occupational Health and Safety (funded by the federal government) and the Association of Workers' Compensation Boards of Canada (funded by provincial

Workers' Compensation Boards) to their meetings. CMEC has directly engaged with the leaders of Aboriginal organizations, inviting them to join ministers for part of their meetings. There is no evidence of these kinds of connections by the FLMM or Social Services ministers. None of the forums have any relationship with committees in the House of Commons or in provincial legislative assemblies. In our executive-dominated Westminster system, not even federal-provincial agreements are brought forward to be ratified by legislators. These practices are long-standing and have not changed under the federal Conservatives.

The most institutionalized "intermediary organizations" in human capital matters are in the education sector. These are financially supported in a variety of ways. First, the federal government provides considerable funding to a large number of research granting organizations. Second, the Canadian Education Statistics Council and the Canadian Information Centre for International Credentials are actually part of CMEC, and funded by governments. In addition, there are a wide variety of organizations funded through memberships that interact with CMEC, even though they do not have a direct relationship. These include the Association of Universities and Colleges of Canada, the Canadian Association of University Teachers, the Canadian Federation of Students, and the Canadian Graduate Council. It is noteworthy that the Association of Universities and Colleges of Canada has a formal agreement with Ottawa on direct federal investment in post-secondary research, but not CMEC. This agreement was initiated by the federal Liberals. However, two post-secondary education–related research organizations that were funded and established by the federal Liberals in the 1990s and 2000s – the Canadian Millennium Scholarship Foundation and the Canadian Council on Learning – were defunded by the Conservatives after they assumed power in 2006 and have since closed. This defunding is consistent with their view of Ottawa not operating in areas of provincial jurisdiction. They would certainly have noted refusal on the part of some provinces to support the activities of these federally funded organizations.

This Conservative view has carried over into the labour market and social services sectors. Over the past few years, many intermediary organizations that previously received federal funding for research, information, knowledge exchange, and information dissemination have also been defunded. The Canadian Labour and Business Centre closed in 2006, Canadian Policy Research Networks closed in 2009, the Canadian Council on Social Development was defunded in 2011, and the National Council on Welfare closed in 2012. National Sector Councils (37 of them), established to bring together business, labour, and educational stakeholders to share ideas and perspectives about human resources and skills issues, lost their core federal funding in 2013 and many are expected to close. The main intermediary organizations that continue with federal funding operating in human resources matters outside of education are the Canadian Labour Market and Skills Researcher Network out of the University of British Columbia, and the Childcare Resource and Research Unit at the University of Toronto.

The closure of these kinds of institutions has consequences; for example, with the demise of the Canada Assistance Plan in 1996, Canada no longer even collects and publishes up-to-date social assistance statistics on a pan-Canadian basis.[7] As a result, the media and the public have no way to compare how provincial governments are managing this very expensive social program, a key barometer of the country's economic well-being. Intermediary organizations funded by membership or foundations such as the Caledon Institute for Social Policy, the Canadian Centre for Policy Alternatives, the Canadian Education and Research Institute for Counselling, the Canadian Association for Supported Employment, and Youth Centres Canada try to fill the gap. However, unlike education sector actors, they do not have access to government data, limiting their analytical capacity.

Outcomes/Outputs

Assessing the ability of federal and provincial governments in Canada to work together to address public policy problems is the most challenging part of this chapter. For this I have looked for public reports and suggestions from those interviewed.

On the labour and CAALL file, there is very limited information available on how governments are working together to address issues such as work-life balance, vulnerable workers, workplace hazardous material, or healthy workplaces. This has not changed since the Conservatives assumed power; very little information was available when the federal Liberals were in charge. The projects outlined on the website are dated, some going back to 2006. The following examples of successful federal-provincial cooperation were identified to me: coordination during the H1N1 pandemic, Canada and the International Labour Organization (ILO), the harmonization of the Workplace Hazardous Materials Information System (WHMIS), and the Young Workers Social Media Initiative.

Public policy challenges for CMEC include recognition and portability of Canadian and international educational and occupational qualifications; K–12 and post-secondary education quality and effectiveness; access to post-secondary education, including affordability through student financial assistance; increasing the attractiveness of Canada to international students; and student and teacher mobility. There is no doubt that Canada is regarded as a success in the Program for International Student Assessment (PISA) studies done by the Organisation for Economic Co-operation and Development (OECD) that compare educational achievements of 15-year-olds, a key measure of effectiveness with respect to K–12

---

[7] On a voluntary basis, with the assistance of the federal government, provincial/territorial directors of Income Support develop and release comparable social assistance statistics. Their latest report was for 2008.

education.[8] CMEC has also played a significant role in addressing the transferability of foreign credentials, as well as increasing the attractiveness of foreign students to Canada through the Education in Canada brand. Both of these successes had the involvement of federal officials. However, there is no evidence that issues related to high student tuition fees are being worked on collectively across governments, despite the increased level of student debt highlighted in the media in 2012. The Canadian Council on Learning (2011) was highly critical of the lack of pan-Canadian progress on many post-secondary issues. When asked about CMEC effectiveness, more than one provincial official suggested that its activities involve too much time and effort for only marginal results.

While the FLMM has made progress in addressing issues related to labour mobility, there are no recent reports publicly available to confirm this. On labour market information, the Advisory Panel on Labour Market Information (2009) was highly critical. Although the federal Conservatives initiated the panel, they have not publicly responded to the panel's recommendations or invested in better information. The FLMM does not seem to engage at all on issues relating to Employment Insurance, where Ottawa continues to make unilateral changes, treating provincial governments like any other stakeholder. Even though the Ontario government-supported Mowat Centre created an entire task force focused on Employment Insurance Reform (Banting and Medow 2012), there is no evidence that recent federal changes responded to their recommendations.

There is also no evidence of governments working together on a pan-Canadian basis through the FLMM to alleviate skills and labour shortages or Aboriginal unemployment. It appears that regional initiatives such as the New West Partnership (between British Columbia, Alberta, and Saskatchewan) are playing a more significant role, not the FLMM. The provincial/territorial portion of the FLMM was certainly successful in pushing back on unilateral changes Ottawa wanted to make through the Canada Job Grant; its effectiveness as a forum for federal/provincial/territorial discussion has still not been proven.

Under the federal Liberals, the Social Services forum had success between 1997 and 2006 in facilitating collaborative federal-provincial work that resulted in improvements to social programs, especially through the National Child Benefit (Finnie and Irvine 2008; National Council on Welfare 2011; Simmons 2008). The Conservative cancellation of the early learning and child-care agreements has been severely criticized by advocates and researchers (Canadian Centre for Policy Alternatives 2011; Mahon and Collier 2010), as well as by provincial governments, and resulted in the disengagement of the federal minister from the work of the Social Services forum. Today there is no shortage of issues that this forum could

---

[8] Some provincial officials suggest that Canada is successful *because* there is no federal minister of education: that it is competition between provinces that accounts for the good results.

be tackling: for example, lack of affordable child care across the country, the rising disability component of social assistance caseloads, income inequality, poverty, and access to employment and living supports for persons with disabilities. While collaborative work may be underway between officials, on a pan-Canadian basis politicians are completely disengaged, with each government figuring out the best approach to meet its individual needs. On disability issues it is noteworthy that in July 2012 the federal Conservatives appointed a panel to consult with private sector employers on the labour market participation of people with disabilities (HRSDC 2013). These efforts did not include provincial participation. While every province (except for British Columbia and Saskatchewan) has developed or is developing a poverty strategy, there is no federal engagement or even an attempt to coordinate and share information on a pan-Canadian basis.

INTERGOVERNMENTAL INSTITUTIONS IN EMPLOYMENT POLICY IN THE EUROPEAN UNION

Comparing Canadian practices to other political systems is useful; by systematic-ally comparing structures and processes, we can shed light on where we have come from, causes and effects, and alternatives to current approaches. In the European Union, like Canada, most social policy matters are the responsibility of the con-stituent units – provinces and territories in Canada and member states in the EU. However, over the past 15 years a significant pan-European dimension to social policy matters has been developed using a technique called the Open Method of Coordination (OMC). This next section provides a brief overview of the institu-tional structure of IGR in the European Union as it applies to employment policy,[9] using the same parameters as in the Canadian part of this chapter. It also highlights important differences.

Although the EU has had a social dimension since its beginning, it was not until 1997 that the Amsterdam Treaty authorized the creation of the European Employment Strategy to combat the challenges all were facing – unemployment, skills and labour shortages, and too many dependent people. The strategy is under the direction of ministers of employment, social protection, consumer protection, health, and equal opportunities from the 28 EU member states that comprise the

---

[9] As in Canada, EU coordination varies by policy sector. The description would be dif-ferent if I were to look at social inclusion or higher education policy. For more detailed information on how the OMC has been used in the EU in social policy matters compared to Canada, see the special issue of *Canadian Public Administration* 56 (June 2013) dedicated to this analysis. See also Wood (2013).

Employment, Social Policy, Health and Consumer Affairs Council (EPSCO). They meet face-to-face around four times a year, supported by the Employment Committee (EMCO), a treaty-based group made up of two director-level subject-expert civil servants per jurisdiction. A network of the Heads of Employment Services (HOPES) from each member state supports EMCO. There are defined processes for EPSCO ministers to meet with social partners and civil society organizations. The European Parliament also provides an opinion on the guidelines underpinning the European Employment Strategy.

Details on the operation of EPSCO and EMCO and their relationship to member-state activities are available on the website of the European Commission's Directorate General for Employment, Social Affairs, and Equal Opportunities, including an annual work plan and results achieved (European Commission 2013). In the EU, intermediary organizations are ubiquitous. Of particular note in employment policy matters are the European Employment Observatory (which provides information, comparative research, and evaluation on employment policies and labour market trends) and the Mutual Learning Program (which helps EU member states learn from each other's experiences and enhance the transferability of good practice). Funded by the Commission (i.e., by the EU member states as part of their contribution to the EU budget), these organizations involve representatives from the EU member states as well as their stakeholders. Intergovernmental agreements are not a big feature of the European Union, except in relation to the European Social Fund.

Why are intergovernmental relations in the EU so different from Canada and much more highly institutionalized? First, heads of state in the EU routinely meet four times a year to plot the overall direction of all EU activities. This contrasts with Canadian first ministers who almost never meet. Second, the EU does not have an equivalent involvement of deputy ministers due to the presence in Brussels of permanent senior level "ambassador-type" member-state representatives who provide quasi-political oversight to the work of officials. Third, while in both Canada and the EU decisions are usually taken by consensus, the EU does have access to qualified majority voting if necessary. With consensus as an operating principle, in Canada issues are often avoided, or decisions are reached that serve the "lowest common denominator." Fourth, the European Commission actually funds some of the costs incurred by EU member states to attend meetings. In Canada travel costs are born by federal and provincial governments, and austerity often means that provincial officials cannot travel to attend face-to-face meetings. Of most significance is the role of the European Commission versus the Government of Canada. Although the European Commission may act in its own interests, it is held in check by a need for member-state agreement to its actions. By its very structure and lack of capacity to provide financial incentives, it can play the "honest broker" role that the Government of Canada often cannot. It also has agenda-setting authority, a role not available to the Government of Canada.

CONCLUSION

This chapter assessed how Canadian governments collectively made the federation work in matters related to human capital development over two time periods: when the Liberals were in power between 1993 and 2005, and under Conservative rule between 2006 and 2012. The focus was on four Canadian intergovernmental forums and related institutions, and one in the European Union. I compared the institutionalization associated with these intergovernmental forums, and the policy capacity in each sector. Table 1 summarizes in chart form the different dimensions of the five intergovernmental institutions reviewed in this chapter.

The Council of Ministers of Education Canada is highly institutionalized based on a variety of factors: a founding agreement, regularized minister and deputy minister engagement, an established pattern of meetings, defined secretariat support, transparency, and participation beyond government executives. In addition, the presence of intermediary organizations provides for research, information, analysis, knowledge exchange, and information dissemination. In many ways CMEC is similar to the significant institutionalization of EPSCO and related organizations in the European Union. Next in terms of institutionalization in Canada is the Canadian Association of Administrators of Labour Legislation, although it is much weaker than CMEC in regard to secretariat support, outputs, transparency, and the involvement of intermediary organizations. With the change of government from Liberal to Conservative in 2006, neither forum has been significantly impacted in its activities; ministers and deputy ministers are still engaged and meet regularly.

This is not the case with the Forum of Labour Market Ministers and Ministers of Social Services. A weak secretariat that rotates every two years between provinces means that the work of the forum is almost completely dependent on the competence, capacity, and interest of the provincial co-chair. Certainly the FLMM is better than MSS in this regard in terms of collective commitment to the work and resources assigned. However, both forums are virtually invisible, as ministers rarely meet and deputy ministers only engage from time to time. Both of these forums have been significantly impacted by the change of government from federal Liberal to Conservative rule. Prime Minister Harper does not meet with premiers on a multilateral basis, and his example has set the tone at the sector level among his ministers. As commentator Bruce Anderson (2013) has observed, "Something important has been lost with the demise of First Ministers' meetings – a sustained discussion on how the different parts of Canada work together.... Without First Ministers' meetings, public engagement with national politics has declined."

More significant to the decline as it relates to these two policy sectors is the "deinstitutionalization" that has occurred due to the defunding of intermediary organizations that used to do research and connect governments and stakeholders in the two sectors on a pan-Canadian basis: the Canadian Labour and Business Centre, Canadian Policy Research Networks, the Canadian Council on Social Development,

Table 1: Comparing Five Intergovernmental Forums in Human Capital Development between 1993 and 2012

Name of Forum	Canadian Association of Administrators of Labour Legislation (CAALL)	Council of Ministers of Education Canada (CMEC)	Forum of Labour Market Ministers in Canada (FLMM)	Ministers of Social Services in Canada (MSS)	Employment, Social Policy, Health and Consumer Affairs Council (EPSCO in EU)
Interdependence	Low	High	High	Medium	Medium
Nature and composition of the forum	F/P/T	P/T	F/P/T	F/P/T	Member states and Commission
	Ministers and deputy ministers	Ministers and deputy ministers	Assistant deputy ministers	Officials	Ministers and officials
	Formal	Formal	Informal	Informal	Treaty based
Secretariat and functioning	Permanent	Permanent	Rotating	Rotating	Rotating
	Federal	Provincial	Provincial	Provincial	Member state
	Consensus	Consensus	Consensus	Consensus	Consensus/ QMV
Relationship to First Ministers	None	Medium	High under Liberals, then medium under Conservatives	High under Liberals, now low under Conservatives	Consistently high
IGR agreements	None	A few	Extensive with 49 agreements	High under Liberals, now low under Conservatives	Minor, except for European Social Fund
Transparency	Website	Website	Website improved under Conservatives	No website; those in place under Liberals discontinued	Many websites
	Limited information	Extensive information	Some information	No information	Extensive information

... continued

Table 1 (Continued)

Name of Forum	Canadian Association of Administrators of Labour Legislation (CAALL)	Council of Ministers of Education Canada (CMEC)	Forum of Labour Market Ministers in Canada (FLMM)	Ministers of Social Services in Canada (MSS)	Employment, Social Policy, Health and Consumer Affairs Council (EPSCO in EU)
Participation beyond government	Some involvement	Some involvement	Low involvement	Low involvement	Extensive involvement of NGOs plus parliaments
Involvement of non-government organizations (NGOs)	NGOs government and member-ship funded	NGOs government and member-ship funded	NGOs defunded under Conservatives	NGOs defunded under Conservatives	NGOs funded by European Commission
Outcomes/ outputs	Low	Medium	Low	High under Liberals, now low under Conservatives	Medium

Note: F/P/T = federal/provincial/territorial; IGR = intergovernmental relations;
P/T = provincial/territorial; QMV = qualified majority voting.

Source: Created by author.

the Canada Millennium Scholarship Foundation, the Canadian Council on Learning, the National Council on Welfare, and National Sector Councils. With the exception of the Canadian Labour and Business Centre, all of these organizations lost government support after the Conservatives assumed power in 2006.

What about policy capacity, that is, the ability of Canadian governments to work together to address public policy problems? Polls in Canada consistently demonstrate that Canadian citizens want their governments to work together to address the issues facing the country regardless of how powers are divided in the federation (Fafard, Rocher, and Cote 2010). Even the Supreme Court of Canada has recently highlighted how constitutional principles and practice in Canada demonstrate that "cooperation [between federal and provincial governments] is the animating force. The federalism principle upon which Canada's constitutional framework rests demands nothing less."[10]

---

[10] Supreme Court of Canada (2011), *Reference re Securities Act*, SCC 66, para. 133.

Despite the wishes of the Canadian public and the Supreme Court of Canada, in my view our current intergovernmental structures in human capital development are not particularly conducive to facilitating constructive dialogue and cooperation between governments. CAALL outputs are slim. CMEC by its very structure has not provided a place for federal/provincial/territorial governments to come together and bridge their differences on post-secondary education matters. In the absence of a federal-provincial structure, intermediary organizations have tried to play a role, but in many ways this has only increased incoherence in post-secondary education, with provinces now covering the cost of teaching and the federal government covering the cost of research. Governments are avoiding key issues such as student debt loads.

Lack of engagement by ministers and deputy ministers plus weak and rotating secretariats in the FLMM and MSS policy domains mean that many policy issues are being ignored or avoided: for example, employment insurance reform, skills and labour shortages, rising numbers of people with disabilities on social assistance, income inequality, and lack of access to affordable child care. In that regard we could look to the past, reflecting on positive developments, like the National Child Benefit, that came about as a result of concerted and coordinated federal-provincial effort. And we could also look to the European Union, reflecting in particular on the techniques that member states use to facilitate positive relations on a voluntary basis.

In the EU the direct costs of engagement, active coordination between governments and sectors, and research and knowledge exchange are all viewed as legitimate costs to support the ongoing management of the union. Likewise, in Canada, we view an investment in CMEC as producing positive K–12 education results. Given interdependence in the labour market and social services sectors, why do we not likewise invest in intergovernmental institutions and related intermediary organizations? In my view, we cannot manage a complicated federation like Canada's on an ad hoc and shoestring basis – institutions matter. As outlined in this chapter, the ones we have in place for human capital development are coping, some better than others, but we would be well served to shine a light on what goes on under the surface through our various intergovernmental structures. Hopefully this chapter has started that process.

REFERENCES

Advisory Panel on Labour Market Information. 2009. *Working Together to Build a Better Labour Market Information System for Canada*. Accessed May 31, 2012. http://publications .gc.ca/collections/collection_2011/rhdcc-hrsdc/HS18-24-2009-eng.pdf.

Agranoff, R. 2004. "Autonomy, Devolution and Intergovernmental Relations." *Regional and Federal Studies* 14 (1): 26-65.

Anderson, Bruce. 2013. "First Ministers Meetings Serve the Long-Term Interests of Canada."
    Globe and Mail, September 16, 2012. Accessed February 28, 2013. http://www.the
    globeandmail.com/news/politics/second-reading/first-ministers-meetings-serve-the-
    long-term-interests-of-canada/article4548361/.
Bakvis, H., and G. Skogstad. 2012. *Canadian Federalism: Performance, Effectiveness and
    Legitimacy*. Don Mills, ON: Oxford University Press.
Banting, Keith. 2005. "Canada Nation Building in a Federal Welfare State." In *Federalism
    and the Welfare State, New World and European Experiences*, edited by Herbert Obinger,
    Stephan Leibfried, and Frank Castles, 89-137. Cambridge: Cambridge University Press.
Banting, Keith, and Jon Medow, eds. 2012. *Making EI Work: Research from the Mowat Centre
    Employment Insurance Task Force*. Kingston and Montreal: Queen's School of Policy
    Studies, McGill-Queen's University Press. Accessed June 18, 2013. http://www.mowat
    eitaskforce.ca/.
Bolleyer, Nicole. 2006. "Federal Dynamics in Canada, the United States and Switzerland:
    How Substates' Internal Organization Affects Intergovernmental Relations." *Publius:
    The Journal of Federalism* 36 (4): 471-502.
CAALL (Canadian Association of Administrators of Labour Legislation). 2013. Resource
    Library. Accessed June 17, 2013. http://www.caall-acalo.org/en/resource-library/overview.
Cameron, David M. 2005. "Collaborative Federalism and Postsecondary Education: Be
    Careful What You Wish For." In *Higher Education in Canada*, edited by Charles M.
    Beach, Robin W. Boadway, and R. Marvin McInnis, 205-27. Montreal and Kingston:
    John Deutsch Institute for the Study of Economic Policy, Queen's University.
Canadian Centre for Policy Alternatives. 2011. *Alternative Federal Budget 2011: Rethink,
    Rebuild and Renew, a Post-Recession Recovery Plan*. Accessed January 17, 2012. http://
    www.policyalternatives.ca/AFB2011.
Canadian Council on Learning. 2011. *What Is the Future of Learning in Canada?* Accessed
    February 18, 2013. http://www.ccl-cca.ca/pdfs/CEOCorner/2010-10-11WhatistheFuture
    ofLearninginCanada.pdf.
CICS (Canadian Intergovernmental Conference Secretariat). 2013. Publications page.
    Accessed September 3, 2013. http://www.scics.gc.ca/english/view.asp?x=198.
CMEC (Council of Ministers of Education Canada). 2003. *Framework for the Future*.
    Accessed February 18, 2013. http://www.cmec.ca/Publications/Lists/Publications/
    Attachments/51/CMECReview.en.pdf.
—. 2013. Publications page. Accessed June 17, 2013. http://www.cmec.ca/9/Publications/
    index.html.
Council of the Federation. 2013. "Canadian Premiers Continue to Have Concerns with
    Proposed Canada Job Grant." News release, July 25, 2013. Accessed September 3, 2013.
    http://www.councilofthefederation.ca/en/latest-news/13-2013/328-canada-s-premiers-
    continue-to-have-concerns-with-proposed-canada-job-grant.
Dennison, Don. 2005. "Intergovernmental Mechanisms: What Have We Learned?"
    Presented at the Looking Backward, Thinking Forward conference for the Institute of
    Intergovernmental Relations, Queen's University, Kingston, May 12–14.

Dupré, Stefan. 1985. "Reflections on the Workability of Executive Federalism." In *Intergovernmental Relations*, edited by Richard Simeon, 1-32. Toronto: University of Toronto Press in cooperation with the Royal Commission on the Economic Union and Development Prospects for Canada.

European Commission. 2013. "Employment, Social Affairs and Exclusion." Accessed June 18, 2013. http://ec.europa.eu/social/search.jsp?langId=en&menuType=basic.

Fafard, Patrick, Francois Rocher, and Catherine Cote. 2010. "The Presence (or Lack Thereof) of a Federal Culture in Canada: The Views of Canadians." *Regional and Federal Studies* 20 (1): 19-43.

Finnie, Ross, and Ian Irvine. 2008. *The Welfare Enigma: Explaining the Dramatic Decline in Canadians' Use of Social Assistance 1993–2005*. C.D. Howe Institute Commentary, No. 267. Accessed May 8, 2013. http://www.cdhowe.org/pdf/commentary_267.pdf.

FLMM (Forum of Labour Market Ministers). 2013. "FLMM – Who We Are." Accessed June 17, 2013. http://www.flmm-fmmt.ca/english/view.asp?x=1.

Haskel, Barbara. 2013. "'Where There's a Will ...': Reforming Postsecondary Education in Canada's and the European Union's Decentralized Systems." *Canadian Public Administration* 56 (2): 304-21.

Hooghe, Liesbet, Gary Marks, and Arjan H. Schakel. 2010. *The Rise of Regional Authority: A Comparative Study of 42 Countries*. London: Routledge.

HRSDC (Human Resources and Skills Development Canada). 2013. *Rethinking DisAbility in the Private Sector.* Report from the Panel on Labour Market Opportunities for Persons with Disabilities. Accessed June 18, 2013. http://www.hrsdc.gc.ca/eng/disability/consultations /rethinking_disabilities.shtml.

Hueglin, Thomas. 2013. "Comparing Canada and the European Union: Treaty Federalism as a Model of Policy Making." *Canadian Public Administration* 56 (2): 185-202.

Inwood, Gregory, Carolyn Johns, and Patricia O'Reilly. 2011. *Intergovernmental Policy Capacity in Canada: Inside the Worlds of Finance, Environment, Trade and Health*. Montreal and Kingston: McGill-Queen's University Press.

Kennett, Steven. 1998. "Securing the Social Union: A Commentary on the Decentralized Approach." Research Paper No. 34. Institute of Intergovernmental Relations, Queen's University, Kingston.

Mahon, Rianne, and Cheryl Collier. 2010. "Navigating the Shoals of Canadian Federalism: Childcare Advocacy." In *Federalism, Feminism and Multilevel Governance*, edited by Melissa Haussman, Marian Sawer, and Jill Vickers, 51-66. Surrey, England: Ashgate.

National Child Benefit. 2013. Home page. Accessed June 17, 2013. http://www.national childbenefit.ca/eng/home.shtml.

National Council on Welfare. 2011. *The Dollars and Sense of Solving Poverty*. Autumn. Accessed October 22, 2014. http://publications.gc.ca/collections/collection_2011/cnb-ncw/HS54-2-2011-eng.pdf.

Poirier, Johanne. 2001. "The Functions of Intergovernmental Agreements: Post Devolution Concordats in a Comparative Perspective." Working Paper, The Constitution Unit, University College London.

Simmons, Julie. 2008. "What Causes Policy Convergence in a Federal Setting? Lessons from the Provincial Components of Canada's National Child Benefit." Prepared for the Canadian Political Science Annual Meeting, June 4–6.

Whittington, Les. 2012. "McGuinty Laments Harper's Refusal to Attend Premiers' Meeting in Halifax." *Toronto Star*, November 20, 2012. Accessed June 17, 2013. http://www.thestar. com/news/canada/2012/11/20/mcguinty_laments_harpers_refusal_to_attend_premiers _meeting_in_halifax.html.

Wood, Donna E. 2013. "Using European Ideas to Open Up Canadian Federalism: The Case of Employment Policy." Research paper prepared for EURAC Institute for Studies of Federalism and Regionalism. Posted online as a European Autonomy and Diversity paper. http://www.eurac.edu/en/research/institutes/imr/activities/bookseries/edap/default.html.

Wood, Donna E., and Thomas Klassen. 2012. "The Governance Problem in Employment and Training Policy in Canada." In *Making EI Work: Research from the Mowat Centre Employment Insurance Task Force*, edited by Keith Banting and Jon Medow, 449-76. Kingston and Montreal: Queen's School of Policy Studies, McGill-Queen's University Press.

8

LE GOUVERNEMENT RÉGIONAL D'EEYOU ISTCHEE-BAIE-JAMES : UNE FORME NOVATRICE DE GOUVERNANCE CONSENSUELLE AU CANADA

Geneviève Motard[1]

INTRODUCTION

Le 24 juillet 2012, le Grand Conseil des Cris (Eeyou Istchee) et le Gouvernement du Québec concluaient une entente visant à réformer les institutions politiques de la région de la Baie James couvrant la majeure partie du territoire traditionnel des Cris de l'est de la Baie James, Eeyou Istchee (Gouvernement du Québec 2012; Entente sur la gouvernance dans le territoire d'Eeyou Istchee Baie-James entre les Cris d'Eeyou Istchee et le Gouvernement du Québec 2012, ci-après Entente; Entente sur certaines questions liées à l'entente sur la gouvernance dans le territoire d'Eeyou Istchee Baie-James entre les Cris d'Eeyou Istchee et le Gouvernement du Québec 2012). Cette entente procède à l'abolition d'institutions existantes, elle attribue de nouveaux pouvoirs à d'autres institutions et crée une nouvelle instance régionale. Le régime de gouvernance cri n'en est

[1] L'auteure souhaite remercier M. Matthieu Juneau, étudiant au doctorat en droit à l'Université Laval, pour le travail de traduction de la conférence présentée initialement ainsi que M. Dave Guénette, étudiant à la maîtrise en études internationales à l'Université Laval, pour le travail de recherche et de révision des notes. L'auteure souhaite également exprimer sa gratitude aux évaluateurs anonymes pour l'ensemble de leurs commentaires constructifs.

pas simplifié pour autant, puisque les anciennes structures locales demeurent en place et qu'une structure régionale politiquement plus complexe que celle qui la précède vient s'y ajouter (Otis et Motard 2009). Le texte qui suit a pour objectifs de présenter la structure juridique de la nouvelle instance régionale et de vérifier la mesure dans laquelle cette institution s'inspire structurellement de la théorie de la démocratie consociative. La démocratie consociative cherche à répondre aux besoins de gouvernance partagée de groupes ethniques fortement divisés, mais coexistant sur un même territoire ou partageant des territoires limitrophes. Le recours à cette forme institutionnelle peut avoir différents objectifs, comme le sont la stabilisation des institutions démocratiques, la pacification des relations intercommunautaires, la réconciliation de communautés fortement divisées ou encore la juste représentation des minorités dans les institutions de gouvernance. Or, au Canada, la progression de la colonisation a eu pour conséquence que, dans bien des régions, les peuples autochtones et les membres de la société majoritaire coexistent aujourd'hui sur un même territoire. Un des objectifs poursuivis dans ces pages est d'entamer une réflexion sur le potentiel de la démocratie consociative à répondre aux demandes d'autonomie des peuples autochtones au Canada. Pour ce faire, nous aborderons d'abord le contexte entourant l'émergence du Gouvernement régional, nous présenterons ensuite la structure mise en place et nous ciblerons, pour finir, les principaux éléments de l'entente de 2012 qui, dans la doctrine sur le consocialisme, ont été identifiés comme étant potentiellement problématiques, car créant des situations d'échec ou d'instabilité.

PARTIE I – LE CONTEXTE ENTOURANT L'ÉMERGENCE DU GOUVERNEMENT RÉGIONAL

La conclusion de l'entente à l'origine du gouvernement régional s'inscrit dans un contexte politique et économique particulier. Plusieurs motifs économiques et politiques expliquent en effet que les autorités cries et québécoises en soient ici arrivées à une entente. La raison principale derrière la réforme actuelle se trouve dans le conflit qui a découlé de la politique de décentralisation ou « politique de gouvernance régionale » mise en place au début des années 2000 par le gouvernement du Québec. Alors que, de leur côté, les Cris de l'est de la Baie James réclament la réforme de leurs structures de gouvernance depuis quelques années déjà, les mésententes entre le Grand Conseil des Cris (Eeyou Istchee) et les municipalités concernant leurs rôles respectifs sur le territoire et notamment les différends concernant le rôle de la Conférence régionale des Élus de la Baie-James (CRÉ-BJ) et du Fonds de développement régional avaient exacerbé cette situation et mené à la nomination d'un conciliateur en 2008 (Gouvernement du Québec 2009, 68). Le professeur Grammond observe que, de manière générale, les structures issues de la régionalisation entreprise au Québec depuis le début des années 2000 (Grammond

2009, 950-953; Grammond 2008; Morin 2006, 57-58; Glenn 1986; Proulx 2002, 129-190) suscitent la méfiance des communautés autochtones du fait que celles-ci « [s]e trouvent [à l'instar des autres acteurs locaux] à la base [de la gouvernance territoriale], et y occupent au surplus une position très minoritaire. Le contraste est frappant avec la gouvernance partagée, davantage horizontale, où des relations « de nation à nation » s'établissent entre l'État et les autochtones » (Grammond 2009, 952). De manière plus spécifique, la politique de gouvernance régionale posait trois problèmes importants pour les Cris.

La source principale du mécontentement des Cris se trouve dans le texte même de la *Convention de la Baie-James et du Nord québécois* (Convention de la Baie-James et du Nord québécois et conventions complémentaires 2012, ci-après CBJNQ), puisque celle-ci exclut largement les Cris de la gestion d'ensemble du territoire traditionnel cri, Eeyou Istchee. En vertu de la CBJNQ, les Cris ne sont en effet pas responsables de l'administration des terres de catégorie III, lesquelles composent la majorité du territoire traditionnel cri[2]. Depuis la conclusion de la CBJNQ, ces terres sont en effet administrées par la Municipalité de la Baie-James (MBJ) (Loi sur le développement et l'organisation municipale de la région de la Baie James, ci-après LDOMBJ, art. 34), laquelle était dirigée, depuis sa création en 1971, par le conseil d'administration de la Société de Développement de la Baie-James (SDBJ), une société de développement économique dont le conseil est désigné par décret gouvernemental. La MBJ n'était donc pas une entité démocratique. Malgré cela, celle-ci pouvait exercer des pouvoirs de nature municipale. En 2001, le gouvernement du Québec procède à une réforme qui démocratise la direction de la MBJ. Depuis, le conseil de la MBJ est composé des maires des municipa-lités de Chibougamau, de Lebel-sur-Quévillon, de Matagami, de Chapais et des

---

[2] La CBJNQ met en place un régime de terres particulier qui s'applique à l'ensemble du territoire couvert par cette convention. S'agissant des Cris, la CBJNQ crée trois catégories de terres : les terres de catégorie I (qui comprend les terres de catégorie IA, IB, et IB spé-ciales) ainsi que les terres de catégorie II et III. Les terres de catégorie 1A sont mises de côté à l'usage et au bénéfice exclusif des Cris, mais le Québec en conserve la nue-propriété (cl. 5.1.2). Aussi mises de côté pour l'usage et le bénéfice exclusif des Cris, les terres de catégorie IB sont cependant la propriété de corporations provinciales cries (cl. 5.1.3). Ces terres sont limitrophes, elles sont les plus exigües puisqu'elles totalisent environ 5 500 km². Ce sont sur ces catégories de terre que sont situés les villages cris (terres de catégorie IB) et les administrations locales cries (terres de catégorie IA). Les terres de catégorie II et III font partie du domaine de la couronne provinciale, mais les Cris peuvent exercer des droits exclusifs de chasse, pêche, trappage sur les terres de catégorie II et ont aussi l'exclusivité de l'exploitation de certaines espèces sur les terres de catégorie III (voir par exemple : cl. 24.3.26, 24.3.32 et annexe 2 du chapitre 24). Ces dernières sont assujetties aux règles régissant les terres publiques (cl. 5.3.1).

présidents des localités de Villebois, de Valcanton et de Radisson. Les Cris n'ont pas été invités à siéger au conseil de la MBJ du fait que les terres administrées par leurs municipalités ne font pas partie du territoire sous la juridiction de la MBJ (CBJNQ 2012, cl. 5.1.2 et 10.0.2; LDOMBJ, art. 40(2); Loi sur le régime des terres dans les territoires de la Baie-James et du Nouveau-Québec, art. 20). En d'autres termes, bien qu'ils occupent le territoire sous la juridiction de la MBJ, les Cris ont largement été écartés de la gestion de ce territoire, au contraire des maires des autres municipalités et des présidents des localités de la région. Or, une partie des terres de catégorie I des Cris sont gouvernées par des maires de municipalités, au même titre que le sont les autres municipalités de la région et, à cet égard, nous estimons que rien ne justifiait leur exclusion du conseil de la MBJ.

L'insuffisance de la représentation des Cris dans les institutions chargées de la gestion des terres de catégorie III, savoir la Municipalité de la Baie-James (MBJ) et la Société de Développement de la Baie-James (SDBJ), alors que les Cris représentent plus de soixante pour cent de la population vivant sur le territoire, que leurs territoires familiaux sont couverts par cette catégorie de terre et qu'ils sont dirigés par des maires au même titre que les autres municipalités et localités qui siègent à la MBJ a été un facteur important de litige entre le gouvernement du Québec et les Cris (Otis et Motard 2009, 140; Paul 2010, 3).

Le deuxième problème à la base du différend ayant mené à la création du gouvernement régional repose sur l'inefficacité du Conseil régional de zone de la Baie James mis en place par la CBJNQ, sur lequel siègent les Cris à parité avec les représentants de la MBJ et qui a pour mandat de gérer les terres de Catégorie II sur lesquelles les Cris disposent de droits de chasse, de pêche et de trappage exclusifs (CBJNQ 2012, ch. 11B, cl. 11B.0.16 et ch. 24). S'il faut bien voir que ce Conseil n'a, « à toute fin pratique, jamais fonctionné », la cause en est le financement inadéquat qui lui a été réservé au fil des ans, ce qui ne lui a pas permis de jouer son rôle (Paul 2010, 2). Mais encore, en vertu de la *Loi sur le Conseil régional de zone de la Baie James*, la SDBJ qui constituait jusqu'en 2001 le Conseil d'administration de la MBJ, nommait aussi les membres siégeant sur le Conseil régional de zone de la Baie James. Comme l'explique le juge Réjean Paul dans ses rapports de 2008 et de 2010, la *Loi sur le développement de la Baie James et d'autres dispositions législatives* a eu pour effet de transférer le pouvoir de nomination de la SDBJ à la MBJ, laquelle se voyait dès lors accroître son pouvoir de gestion sur les terres de catégorie II en plus de son pouvoir d'administration sur les terres de catégorie III (Paul 2010, 2). Autrement dit, alors que les Cris ne voient pas augmenter leur pouvoir de gouverner les terres de catégorie III dont la gouvernance est réservée exclusivement à la MBJ, les élus des villes avoisinantes se voient en revanche augmenter leur pouvoir à l'égard des terres de catégorie II. Les modifications législatives ayant eu lieu sans consulter ni tenir compte de l'avis des Cris, il n'est pas étonnant que cette situation ait été source de litige avec le gouvernement du Québec.

Enfin, le dernier problème à l'origine du différend ayant mené à la création du gouvernement régional concerne la mise sur pied, par les autorités québécoises, de la CRÉ-BJ. Les CRÉ ont, de façon générale :

> « [l]e mandat d'« évaluer les organismes de planification et de développement locaux et régionaux », de « favoriser la concertation des partenaires de la région », d'« établir un plan quinquennal de développement » et de « donner, le cas échéant, des avis au ministre sur le développement de la région » » (Morin 2006, 46; Gouvernement du Québec 2004, 29; Loi sur le ministère des Affaires municipales, des Régions et de l'Occupation du territoire, ci-après LMAMROT, art. 21.7).

Ce faisant, ils deviennent le principal interlocuteur du gouvernement du Québec en ce qui concerne le développement régional (LMAMROT, art. 21.6) et décident parallèlement du financement des projets de développement qu'ils jugent prioritaires (LMAMROT, art. 21.18 et 21.23.1). Ce sont en effet les CRÉ qui décident de la répartition des montants composant le Fonds de développement régional. Or, la CRÉ-BJ est composée du maire de la MBJ, des maires des villes de Chapais, de Chibougamau, de Lebel-sur-Quévillon et de Matagami (LMAMROT, art. 21.5). L'Administration Régionale Crie agit quant à elle à titre de CRÉ pour les communautés cries (LMAMROT, art. 21.5). À l'instar de la situation qui prévaut au sein de la MBJ, les Cris ont par conséquent aussi été exclus de cette structure de gouvernance régionale, ce qui signifie qu'ils ne participent ni aux décisions relatives aux priorités de développement qui concernent plus largement Eeyou Istchee/Baie-James ni aux décisions de financement qui y sont rattachées.

En somme, malgré la conclusion, en 2002, de la *Paix des Braves* entre le Grand Conseil des Cris (Eeyou Istchee) et le gouvernement du Québec (Entente concernant une nouvelle relation entre le gouvernement du Québec et les Cris du Québec 2002), les autorités québécoises ont entrepris au tournant des années 2000 le réaménagement des structures de gestion territoriale sans procéder à une consultation spécifique et prioritaire des Cris (Paul 2010, 3; *R. c. Sparrow* 1990; *R. c. Badger* 1996; *Delgamuukw* c. *Colombie-Britannique* 1997; *Nation Haïda* c. *Colombie-Britannique (Ministre des Forêts)* 2004, paragr. 32; *Première nation Tlingit de Taku River* c. *Colombie-Britannique (Directeur d'évaluation de projet)* 2004; *Première nation crie Mikisew* c. *Canada (Ministre du Patrimoine canadien)* 2005; *Beckman* c. *Premières nations Little Salmon/Carmack* 2010, paragr. 61), lesquelles avaient de surcroît pour effet de renforcer la participation des élus locaux – à l'exception des Cris – à la gouvernance régionale (The Grand Council of the Crees 2011, 20; Iserhoff 2011). À notre sens, le principe de l'honneur de la Couronne exigeait qu'une telle consultation soit tenue.

L'ensemble de cette situation a amené le juge Réjean F. Paul, nommé à titre de conciliateur, à recommander, dans son rapport rendu le 4 novembre 2008, que :

> (i) [l]es Cris d'Eeyou Istchee devaient être engagés dans la gestion de l'ensemble du territoire visé par la CBJNQ, incluant les terres des catégories II et III sur lesquelles

sont situés certains « terrains de trappage » (ou territoires familiaux traditionnels des Cris), et (ii) [l]es Cris devaient participer activement à l'établissement d'un régime municipal moderne, au sein duquel ils auront leur place légitime dans la gouvernance de leur territoire visé par la CBJNQ (Paul 2008).

La nécessité de réformer les institutions de gouvernance de ce territoire ressort aussi d'un contexte plus général, savoir que le Grand Conseil des Cris (Eeyou Istchee) s'était par ailleurs entendu, en 2008, avec les autorités fédérales pour procéder à une réforme institutionnelle (Entente concernant une nouvelle relation entre le gouvernement du Canada et les Cris d'Eeyou Istchee 2008). À cet égard, la complexité des institutions locales actuelles qui, pour une même communauté crie, oblige l'adoption de la réglementation en double, nécessite, à sa face même, une réforme des plus urgente (Entente 2012, cl. 8; Loi sur les villages cris et le village naskapi; Loi sur les Cris et les Naskapis du Québec; Murdoch 2008; Otis et Motard 2009, 136). En outre, les conflits découlant de l'inefficacité du contrôle des activités des usagers allochtones en territoire cri constituent aussi une source de problèmes pour les Cris depuis de nombreuses années (Scott et Webber 2001). Finalement, les développements des dernières années concernant la reconnaissance du droit à l'autonomie par les autorités fédérales dans le *Guide de la politique fédérale sur l'autonomie gouvernementale* (Gouvernement du Canada 1995) et le nombre croissant d'ententes entre d'autres nations autochtones et les autorités territoriales, provinciales et fédérales ont sans doute contribué à favoriser l'émergence de nouvelles institutions en territoire cri (Accord définitif de la Première nation Tsawwassen 2007; Accord définitif Nisga'a 1999; Accord sur des revendications territoriales entre les Inuit du Labrador et Sa Majesté La Reine du Chef de Terre-Neuve-et-Labrador et Sa Majesté La Reine du Chef du Canada 2005; Accord sur l'autonomie gouvernementale de la Première nation de Westbank entre Sa Majesté la Reine du chef du Canada et la Première nation de Westbank 2003; Accord sur les revendications territoriales et l'autonomie gouvernementale entre le peuple Tlicho et le Gouvernement des Territoires du Nord-Ouest et le Gouvernement du Canada 2003).

Ajoutons que les projets d'exploitation des ressources du territoire nordique – dont une partie est couverte par la CBJNQ - qui sont promus par le gouvernement du Québec sous le nom de « Plan Nord » ont sans aucun doute favorisé la mise en place de conditions favorables à la signature de cette entente. En effet, il faut d'abord compter sur le fait que les tribunaux canadiens obligent les autorités gouvernementales à consulter, voire à accommoder, les populations autochtones lorsque des mesures portent atteinte à leurs droits. La décision rendue par la Cour suprême du Canada dans l'affaire *Beckman c Première nation Little Salmon/Carmacks* confirme l'obligation des gouvernements d'agir honorablement et de consulter les populations autochtones même en l'absence de dispositions expresses à cet effet dans le texte d'un accord de revendication territoriale globale (*Beckman* c. *Premières nations Little Salmon/Carmack* 2010, paragr. 61). En outre, la *Déclaration des Nations*

Unies sur les droits des peuples autochtones (2007), appuyée par le Canada le 12 novembre 2010, prévoit notamment l'obligation des États d'obtenir le consentement libre et éclairé des peuples autochtones avant d'entreprendre l'exploitation de leurs terres ancestrales (art. 19, 28 et 32). Or, les développements industriels envisagés par le gouvernement du Québec dans le Plan Nord sont susceptibles de porter atteinte aux droits des Cris reconnus dans la CBJNQ, mais aussi à leur mode de vie compte tenu du fait que de nombreux territoires familiaux traditionnels se situent hors des zones d'exploitation exclusive, soit hors des terres de catégorie I et II (The Grand Council of the Crees 2011, 18). L'accent mis par les autorités gouvernementales sur le développement minier renforce cette appréhension, ce qui s'explique fort légitimement par les craintes suscitées par cette forme d'exploitation, comme elle s'est exercée historiquement (Ritter 2001, 227-230)[3]. On peut aussi craindre la pression de l'arrivée de nombreux travailleurs du sud sur la faune ainsi que les effets du « fly-in, fly-out » sur les communautés locales, comme le sont les augmentations appréhendées des taux de criminalité, d'utilisation des drogues, de la prostitution et des jeux de hasard dans les communautés qui seraient situées près des lieux d'exploitation (Storey 2010, 1165; Ritter 2001, 228 et 230-232). En outre, la construction de routes et de nouvelles infrastructures risque notamment de causer des déplacements fauniques de même que de rendre le territoire encore plus accessible, notamment aux touristes, ce qui a le potentiel de porter atteinte aux droits des Cris reconnus dans la CBJNQ (*R. c. Sparrow* 1990, *R. c. Badger* 1996, *Rio Tinto Alcan Inc. c. Conseil tribal Carrier Sekani* 2010; Frouin 2001; Le comité organisateur 2012, 107-08; The Grand Council of the Crees 2011, 24 et 42).

C'est pourquoi l'exploitation des terres de catégorie III ne pouvait plus, politiquement et juridiquement, se faire sans la contribution significative des Cris et ce, même si cette prérogative avait été reconnue au gouvernement du Québec dans la CBJNQ. En effet, dans la mesure où le développement des terres de catégorie III a le potentiel de porter atteinte aux droits reconnus aux Cris sur les terres de catégorie I, II et III les autorités québécoises avaient à notre avis, suivant l'affaire *Beckman*, l'obligation constitutionnelle de consulter les Cris (CBJNQ 2012, cl. 5.5.1; *Beckman c. Premières nations Little Salmon/Carmack* 2010, paragr. 61). Or, en plus de régler le différend concernant le rôle des institutions municipales et régionales sur les terres de catégorie III, la création du Gouvernement régional prévient un nombre important de conflits, lesquels devenaient inévitables en raison des projets envisagés par le gouvernement du Québec en territoire cri. À cet égard, on peut en effet constater les divergences se rapportant au développement du territoire dans la réponse qu'ont présentée les Cris au Plan Nord du gouvernement du Québec (The Grand Council of the Crees 2011). Parmi celles-ci, on peut remarquer que les Cris font de la réforme de la gouvernance une condition *sine qua*

---

[3] L'entente prévoit même une présomption d'incompatibilité entre les activités minières et les activités culturelles et traditionnelles cries : entente 2012, cl 51(2).

non du développement économique du territoire : « [w]ithout a governance regime acceptable to the Cree, there will [be] no Plan Nord in Eeyou Istchee » (The Grand Council of the Crees 2011, 12 et 21). En outre, ils souhaitent être impliqués dans la définition des concepts auxquels le gouvernement du Québec a recours dans le Plan Nord et notamment dans la définition de ce que constituent des « activités industrielles » et des « aires protégées » (The Grand Council of the Crees 2011, 13, 34-39, 45, 52-55, 64, 70, 76, 78, 92, 95-102 et 106). Compte tenu du nombre important de conflits susceptibles de se poser et de la nécessité, pour le Québec, de limiter les conflits – et les poursuites judiciaires qui s'ensuivraient (Entente 2012, cl. 205-206; *Grand Chef Matthew Coon Come et al. c. Hydro-Québec, le Procureur général du Québec et le Procureur général du Canada*; *Grand Chef Matthew Coon Come et al. c. Hydro-Québec, le Procureur général du Québec et le Procureur général du Canada*) – en vue notamment d'attirer les investisseurs, la réforme de la gouvernance régionale à Eeyou Istchee/Baie-James devenait un objectif commun aux autorités politiques cries et québécoises.

PARTIE II – LES ATTRIBUTS DISTINCTIFS DU GOUVERNEMENT RÉGIONAL

Les négociations ayant mené à la mise sur pied du gouvernement régional duraient depuis déjà deux ans, soit depuis 2010 (Accord-cadre entre les Cris d'Eeyou Istchee et le Gouvernement du Québec sur la Gouvernance dans le territoire d'Eeyou Istchee Baie-James 2011, préambule). L'entente intervenue en 2012 ne modifie pas toutes les structures de gouvernance, mais constitue une réforme significative des institutions en place.

Le Gouvernement de la nation crie assumera désormais l'administration des terres de catégorie II en ce qui a trait aux affaires municipales de même qu'en ce qui concerne la gestion des terres et des ressources (Entente 2012, cl. 13-22). Pour ce faire, l'entente abolit le Conseil régional de zone de la Baie James et transfère tous ses droits, ses fonctions, ses biens et son passif au Gouvernement de la nation crie (Entente 2012, cl. 23). Celui-ci agira par ailleurs à titre de CRÉ pour les terres de catégorie I et II et aura par conséquent les mêmes pouvoirs que les CRÉ et les Commissions régionales des ressources naturelles et du territoire (ci-après CRRNT), lesquelles sont notamment responsables de l'élaboration des plans régionaux de développement.

Le processus de décision retenu dans l'entente se caractérise par le dialogue et la négociation entre les Cris et les autorités ministérielles québécoises. Ainsi, s'agissant de la gestion et de la planification de l'aménagement des terres et des ressources, il appartient tout d'abord à la Commission crie d'élaborer un « Projet de plan », lequel doit être soumis au Gouvernement de la nation crie qui doit ensuite l'accepter ou y proposer des modifications. Après acceptation, le Gouvernement de la nation crie doit le faire parvenir au Ministère des Ressources naturelles et de la Faune (MRNF) qui doit aussi l'accepter ou en proposer des modifications. Ce n'est qu'après autorisation par le ministre que le « Projet de plan » devient un plan d'aménagement. Du point de vue des revendications d'autonomie des Cris, il s'agit certes d'une amélioration par rapport à la situation précédente – on passe en effet d'une institution paritaire Crie-MBJ à une institution totalement contrôlée par les Cris. Il demeure que les Cris se voient reconnaître ici un pouvoir de participation aux décisions se rapportant à l'aménagement du territoire et non un pouvoir décisionnel, lequel demeure celui du ministre. En effet, bien que la formule choisie élimine le problème de la participation accrue des municipalités avoisinantes à la gouvernance des terres de catégorie II, l'autorité crie demeure en revanche assujettie aux décisions des instances centrales du MRNF (Entente 2012, cl. 27).

S'agissant cette fois de la gouvernance des terres de catégorie III, l'entente abolit la MBJ, celle-ci étant remplacée par le Gouvernement régional Eeyou Ischtee-Baie-James, un organise municipal (Entente 2012, cl. 76). À l'instar de ce qui prévalait sous le régime de la MBJ, la juridiction territoriale du Gouvernement régional exclut les terres de catégorie I, mais cette fois, les terres de catégorie II sont aussi expressément exclues (Entente 2012, cl. 77-78). Le Gouvernement régional aura, selon l'entente, compétence en ce qui concerne la gestion des affaires municipales, ce qui comprend l'ensemble des pouvoirs préalablement exercés par la MBJ, de même que les pouvoirs prévus par la *Loi sur les cités et villes* et la *Loi sur les compétences municipales*. De plus, le Gouvernement régional exercera les mêmes fonctions que celles exercées par les CRÉ – sous réserve des pouvoirs qui continueront à être exercés par la CRÉ-BJ dont les fonctions seront exercées uniquement à l'égard des Jamésiens – et par les CRRNT. Le Gouvernement régional pourra également déclarer exercer les mêmes fonctions qu'une Municipalité régionale de comté (ci-après MRC) (Entente 2012, cl. 123 et s.). Enfin, le Gouvernement régional sera dirigé par un conseil dont les règles d'opération sont prévues à même l'entente; les règles supplétives étant celles prévues par la *Loi sur les cités et villes* du Québec (entente 2012, cl. 90). Force est ici de constater que les institutions prévues à l'entente s'inspirent uniquement des institutions québécoises et non des traditions politiques cries.

Le Gouvernement régional se distingue de toutes les autres institutions existantes à ce jour au Canada. La caractéristique la plus originale du Gouvernement régional ressort sans nul doute de la composition de son conseil et du processus décisionnel qui y aura cours. Le conseil du Gouvernement régional sera ainsi composé, pour les dix premières années de son existence, d'une représentation paritaire des Cris et des Jamésiens. Cet équilibre est appelé à se modifier au terme des dix premières années d'existence du Gouvernement régional, puisqu'après cette période initiale, la représentation de chaque composante sera fonction de la population résidente suivant une formule qui restera à choisir. Dans tous les cas, les représentants des Cris seront désignés par les Cris parmi les élus des Cris, tandis que les représentants des Jamésiens seront choisis parmi les élus des Jamésiens par le ministre des Affaires municipales, des Régions et de l'Occupation du territoire parmi les membres des conseils des Municipalités qui résident sur le Territoire. Outre la répartition équitable des vingt-deux (22) sièges qui composent le Conseil du Gouvernement régional, l'entente ne prévoit pas les méthodes de désignation qui seront utilisées pour choisir les représentants des Jamésiens et des Cris. Concernant les Cris, ce choix nous semble judicieux dans la mesure où l'on respecte ainsi leur droit à l'autodétermination. Ensuite, la répartition des vingt-deux (22) voix reconnues aux Jamésiens sera déterminée par le ministre. Pour ce faire, le ministre tiendra compte du poids démographique de chaque municipalité (entente 2012, cl. 83). La répartition des vingt-deux (22) voix des Cris entre les communautés cries n'est pas prévue par l'entente, ce qui permet encore une fois de respecter le droit à l'autodétermination des Cris.

À la lecture de ces dispositions, qui mettent l'accent sur la parité de représentation, sur l'équilibre dans l'exercice du pouvoir décisionnel et ultimement sur la négociation des normes, il nous est apparu que la forme choisie pour mettre en place le Gouvernement régional s'inspirait de la démocratie consociative. La section suivante s'attache par conséquent à vérifier dans quelle mesure les négociateurs ont eu recours à ce modèle et à cibler les éléments de l'entente qui nous semblent problématiques.

PARTIE III – UNE FORME NOVATRICE DE GOUVERNANCE CONSENSUELLE AU CANADA

La démocratie consociative est une forme de gouvernance basée sur le consensus entre tous les segments d'une société multiethnique ou multinationale en vue d'assurer la coexistence pacifique, la stabilité démocratique et, en fin de compte, une bonne gouvernance : « [t]he essential characteristic of consociational democracy is not so much any particular institutional arrangement as the deliberate joint effort by the elites to stabilize the system » (Lijphart 1969, 213). En ce sens, nous pensons que le modèle institutionnel retenu dans l'entente est une forme de consocialisme. Or, ce modèle n'a jusqu'à maintenant pas été appliqué de manière aussi approfondie dans le contexte de la mise en œuvre du droit inhérent à l'autonomie des peuples autochtones au Canada. En effet, si elle se distingue de toutes les autres institutions canadiennes par sa nature fortement consociative, force est de constater que la forme institutionnelle choisie dans l'entente de 2012 s'inspire des mécanismes de cogestion mis en place dans plusieurs régions du Canada (Rodon 2003). À la différence toutefois de ces comités de cogestion qui s'intéressent à la gestion des ressources de la terre et qui ont généralement un rôle consultatif auprès des ministres, le Gouvernement régional s'est vu reconnaître, comme cela a été exposé précédemment, des pouvoirs consultatifs en matière de gouvernance régionale, mais aussi de nombreux pouvoirs décisionnels de nature municipale.

Le consocialisme a conséquemment été utilisé de manière beaucoup plus poussée dans le cas du Gouvernement régional qu'il ne l'a été jusqu'ici ailleurs au Canada, du moins dans le contexte de la réforme de la gouvernance autochtone. De plus, le caractère démocratique du Gouvernement régional le distingue des mécanismes de cogestion généralement mis en place, par exemple dans les accords d'autonomie gouvernementale et de revendications territoriales globales. En effet, dans ces ententes, ce sont par exemple des agents de l'État – et non pas des élus – qui y siègent. Dès lors, nous ne pouvons que constater que le Gouvernement régional se démarque des autres formes institutionnelles choisies à ce jour pour mettre en œuvre le droit à l'autonomie et à l'autodétermination des nations autochtones au Canada.

Principal penseur de la démocratie consociative, Arend Lijphart (1969, 1999) demeure encore de nos jours la référence première en la matière. Si les auteurs répertorient plusieurs conditions ou sous-conditions pour déterminer la nature consociative ou non d'une structure institutionnelle (Christensen et Studlar 2006), tous s'entendent pour lui reconnaître quatre (4) attributs fondamentaux (Seaver 2000; Sinardet 2011; Iyer 2007; Lemarchand 2006; Cooley et Pace 2012; Spears 2002). Ainsi, dans les mots de Lijphart, la démocratie consociative :

> (…) can be defined in terms of two primary attributes – grand coalition and segmental autonomy – and two secondary characteristics – proportionality and minority veto. Grand coalition, also called power sharing, means that the political leaders of all of the significant segments of a plural (deeply divided) society govern the country jointly. Segmental autonomy means that the decisionmaking is delegated to the separate segments as much as possible. Proportionality is the basic consociational standard of political representation, civil service appointments, and the allocation of public funds. The veto is a guarantee for minorities that they will not be outvoted by a majority when their vital interests are at stake (1985, 4).

La démocratie consociative s'appuie donc sur une coalition entre les élites des différentes composantes de la société, sur la reconnaissance de l'autonomie interne de chacune de ces composantes ainsi que sur le principe de la proportionnalité et la reconnaissance d'un droit de veto à la minorité. Dans l'entente menant à la création du Gouvernement régional, ces quatre attributs ne sont pas tous rencontrés, même que certaines dispositions sont contraires à l'esprit qui anime cette forme démocratique.

S'agissant du premier attribut, soit l'existence d'une grande coalition entre les élites, on observe que les représentants siégeant au Conseil du Gouvernement régional sont des élus cris et jamésiens. Comme ce sont des élus qui exercent, au final, des responsabilités politiques supplémentaires à leurs responsabilités électives principales, cette façon de faire permet aux autorités jamésiennes et cries de conserver le contrôle sur les domaines de compétence reconnus au Gouvernement régional et d'assurer une harmonie décisionnelle entre les différents paliers gouvernementaux. Le principe de représentation paritaire, entre Cris et Jamésiens, au conseil du Gouvernement régional, reflète sans doute le mieux le caractère consociatif des arrangements institutionnels choisis par les négociateurs.

Le caractère consociatif des arrangements institutionnels se confirme aussi par le mode de sélection du président et du vice-président qui sont désignés en alternance par les Jamésiens et les Cris pour des mandats de deux ans. Le segment responsable de la sélection du président ne désigne pas le vice-président (entente 2012, cl. 101). Le caractère consociatif du conseil ressort aussi de la composition du comité exécutif, celui-ci étant paritaire, sauf en ce qui concerne le président du Conseil du Gouvernement régional qui siège d'office au comité exécutif (entente 2012, cl. 112-113). La reconnaissance du cri et du français en tant que langues principales du Gouvernement régional constitue aussi une mesure conforme à l'esprit qui anime le consocialisme (entente 2012, cl. 108, 110-111), ce qui est toutefois limité par l'absence de reconnaissance du cri comme langue de travail du Gouvernement régional (entente 2012, cl. 109). Enfin, l'entente exige aussi une majorité qualifiée pour prendre plusieurs types de décision, par exemple en ce qui concerne le développement et l'aménagement du territoire ou encore ce qui concerne les institutions, ce qui confirme la structure consociative du Gouvernement régional (entente 2012, cl. 107). En clair, nous pensons que le premier attribut est ici certainement rencontré, puisque la bonne marche du Gouvernement régional dépendra des accords politiques entre les élites politiques cries et québécoises.

Le deuxième attribut de la démocratie consociative, c'est-à-dire la reconnaissance d'une autonomie interne à chaque composante de la société, n'est à notre sens pas respecté, du moins pour l'instant. En effet, d'une part, seuls des pouvoirs de nature municipale sont actuellement reconnus aux autorités cries par la CBJNQ (CBJNQ 2012, ch. 9; Saganash 1993, 88). D'autre part, la portée territoriale de cette autonomie est fort limitée, puisque les pouvoirs reconnus aux Administrations locales cries (ALC) et aux villages cris ne s'exercent que sur les terres de catégories I. En revanche, comme cela a déjà été souligné, l'entente de juillet 2012 reconnaît des pouvoirs accrus au Gouvernement de la nation crie au regard des terres de catégorie II. Au final, on ne peut toutefois pas parler ici de la reconnaissance d'une autonomie interne significative qui prendrait sa source ou qui serait fondée sur la souveraineté inhérente des Cris.

L'on franchira sans doute une nouvelle étape dans la direction d'une reconnaissance plus significative de la souveraineté inhérente des Cris au terme de la conclusion d'une entente sur la gouvernance avec les autorités fédérales (Gouvernement du Canada 1995). Dans l'état actuel des choses, l'autonomie interne reconnue aux Cris continuera à être asymétrique en raison des dispositions constitutionnelles se rapportant au partage du pouvoir dans l'État (*Loi constitutionnelle de 1867*, art. 91 et 92, Partie V de la *Loi constitutionnelle de 1982*) ainsi qu'en raison de l'absence de reconnaissance expresse du droit inhérent à l'autonomie et à l'autodétermination des peuples autochtones dans la loi fondamentale de l'État et au refus de la Cour suprême du Canada d'interpréter, du moins jusqu'à ce jour, l'article 35 de la *Loi constitutionnelle de 1982* en ce sens (Partie II de la *Loi constitutionnelle de 1982*; *R. c. Pamajewon*). Dès lors, on ne pourra se surprendre que la jurisprudence interprète de manière parfois restrictive les pouvoirs cris, ceux-ci étant alors interprétés, à tort selon nous, conformément aux règles du droit administratif étatique en matière de pouvoirs délégués (*Eastmain* c. *Gilpin*, 1987, 1644, *Bande de Mistissini* c. *Iserhoff et Conishish-Coon*, 1996, 6). En somme, l'autonomie reconnue à chaque composante de la société demeure toujours très asymétrique, cette asymétrie étant à l'heure actuelle défavorable aux Cris.

La reconnaissance et la mise en place du principe de proportionnalité dans les institutions et l'administration publique constituent le troisième attribut énuméré par Lijphart. Bien qu'il ne s'agisse pas là d'un des attributs principaux, l'exigence de proportionnalité a pour but de favoriser une juste représentation politique des différents groupes d'une société multiethnique ou multinationale. En revanche, l'exigence de proportionnalité est critiquée en présence d'un déséquilibre démographique important entre les groupes, ce qui ne nous paraît pas être le cas ici. À tout prendre, cette exigence n'est, dans l'entente, pas rencontrée. D'abord, la distribution proportionnelle des fonds publics, des emplois et des charges publiques n'est pas garantie par l'entente de juillet 2012. À cet égard, on doit toutefois observer certaines dispositions qui vont dans le sens d'une représentation proportionnelle, comme c'est le cas de la mention selon laquelle le conseil doit assurer « …dans la mesure du possible, un équilibre dans la représentation » aux postes de direction (entente 2012, cl. 118). Ensuite, la représentation des différentes communautés n'est pour l'instant pas proportionnelle, puisque le principe de la parité a été retenu pour les dix (10) premières années du Gouvernement régional. Puisque les Cris représentent plus de cinquante pour cent des personnes qui résident sur le territoire, cela signifie que les Cris sont pour l'heure sous-représentés. De plus, sauf exception, l'entente exclut du calcul menant à établir la population crie, les personnes vivant à l'extérieur des terres visées par la CBJNQ. Une telle clause a pour effet de sous-estimer la composition de la population crie. Cela dit, en vertu de l'entente de juillet 2012, la distribution des sièges et des voix au sein du Gouvernement régional sera, au terme d'une échéance de dix (10) ans, établie sur la base d'une nouvelle formule (entente, cl. 82, 85). Dès lors, pour assurer une représentativité proportionnelle adéquate, il conviendra de tenir compte de la diaspora crie.

Finalement, en ce qui concerne le dernier attribut, soit la reconnaissance d'un droit de veto au groupe minoritaire, celui-ci est nécessaire pour garantir au groupe minoritaire ou vulnérable que l'on ne portera pas atteinte à ses intérêts fondamentaux ou vitaux. Dans l'entente de juillet 2012, on peut avancer qu'un tel droit de veto a été reconnu dans la mesure où une égalité des voix ne permet pas de prendre une décision; le vote étant alors réputé être négatif (entente 2012, cl. 102(3)). En présence de blocage, un mécanisme d'un tout autre ordre a été prévu en vue de résoudre un conflit politique qui perdurerait. Ce mécanisme nous semble contraire au principe du consocialisme, puisqu'il a pour effet non pas de protéger les intérêts vitaux des Cris, mais permet au contraire d'y passer outre. En effet, le mécanisme permet la mise en tutelle du Gouvernement régional si un blocage survient et dure plus de trente (30) jours (entente, cl. 103). Cette décision revient à la Commission municipale du Québec qui peut décréter la mise en tutelle en se basant sur le critère de l'intérêt public. Considérant la nature vague d'un tel critère et la portée attentatoire au droit inhérent à l'autonomie des Cris de la mise en tutelle, cette mesure nous paraît fort critiquable. La décision doit ultimement être confirmée par la Cour supérieure du Québec, ce qui assure une certaine protection pour les Cris[4]. À tout prendre, l'exercice du pouvoir de tutelle par la Commission municipale du Québec devra alors, selon nous, respecter le principe de l'honneur de la Couronne développé dans la jurisprudence canadienne. En d'autres termes, une mise en tutelle ne saurait, à notre sens, être décrétée par les autorités administratives québécoises sans consultation, voire sans approbation, par les Cris. Or, l'entente ne prévoit pas une telle garantie pour les Cris.

La structure de gouvernance établie par l'entente de juillet 2012 est originale à plusieurs égards. Elle se distingue de toutes les autres formes institutionnelles reconnues dans le cadre de la mise en œuvre de la politique fédérale sur l'autonomie gouvernementale et des revendications territoriales globales, tant en matière de cogestion qu'en matière de gouvernance autonome. Il s'agit en fait d'un modèle inspiré à la fois des mécanismes de cogestion dans sa composition, des formes municipales et régionales québécoises en ce qui concerne la portée de ses pouvoirs, mais aussi des conseils de bandes en ce qui concerne son mécanisme de mise en tutelle. En outre, l'analyse des dispositions de l'entente instituant le Gouvernement régional met en lumière certains éléments potentiellement problématiques, car susceptibles d'exacerber des situations conflictuelles. À notre sens, ces éléments sont la sous-représentation des Cris, l'asymétrie dans la reconnaissance de l'autonomie interne et la possibilité de mise en tutelle du Gouvernement régional par une instance québécoise.

CONCLUSION

La démocratie consociative table sur la coopération des élites et sur leurs efforts délibérés à stabiliser le système décisionnel. Les auteurs sont cependant nombreux à dénoncer l'incapacité du modèle consociatif à atteindre cet objectif

---

[4] La Commission municipale du Québec est un organisme administratif, d'enquête et de conseil, spécialisé en matière municipale. Cet organisme agit aussi en tant que tribunal administratif. Sur les pouvoirs de mise en tutelle, voir la *Loi sur la Commission municipale*, LRQ c C-35, art. 38-60.

(Sinardet, 2011; Younis 2011; Spears 2002; Cooley et Pace 2012; Seaver 2000; Garry 2009). Bien que plusieurs facteurs de nature politique et sociale puissent expliquer les succès ou les échecs constatés par ces auteurs, tous s'entendent pour dire que la coopération des élites constitue la clef du succès de ce modèle (Lijphart 1969, 211-212). Dans ce contexte, la contribution du droit demeure limitée. Ainsi, du point de vue juridique, une des difficultés des mécanismes de gouvernance consensuelle se trouve dans l'absence de recours judiciaires en cas de manquement à l'esprit consociatif (ex. : manquement au respect du principe de proportionnalité ou d'autonomie ou encore absence de volonté politique de s'entendre sur une question délicate), aux garanties reconnues, aux arrangements institutionnels ainsi qu'en cas de blocages politiques.

S'agissant de la situation des relations entre les Cris de l'est de la Baie James et le Québec, on ne peut que constater que plusieurs des facteurs favorisant la réussite du modèle consociatif ne sont pas présents. À cet égard, il convient en effet de souligner que les régimes de démocratie consociative auront davantage de chances de succès à stabiliser les rapports politiques lorsque les facteurs suivants seront en place : la durée pendant laquelle l'arrangement consociatif a été en place, la volonté des élites de passer d'une culture politique de compétition à une culture politique de coopération, la présence de menaces externes, le caractère multiple (et non duel) de la balance des pouvoirs, la quantité limitée de responsabilités attribuées à l'instance gouvernementale (Lijphart 1969, 216-219). Dans le cas de l'arrangement institutionnel prévu par l'entente de juillet 2012, un facteur important jouant en faveur de son succès réside dans le fait que les Cris et le Québec n'en sont pas à leur première expérience de gouvernance partagée, bien au contraire (Feit 2009). En revanche, le facteur le plus susceptible de nuire à son succès demeure, selon nous, le fait que le pouvoir sera ici partagé entre seulement deux composantes sociétales unies par un rapport inégalitaire. À partir d'études sur les systèmes de cogestion, les auteurs ont déjà bien montré que les rapports inégalitaires entre l'État et les nations autochtones se perpétuent au sein des instances de coopération inter-gouvernementale (Nadasdy 2003). Or, les règles de fonctionnement du Conseil et les compétences reconnues au Gouvernement régional montrent également que le modèle québécois de gouvernance municipale et régionale a constitué le modèle dominant de référence. La capacité que possède la Commission municipale du Québec de mettre le Gouvernement régional sous tutelle tend à valider ce point de vue. Malgré le fait que ces dispositions puissent être attentatoires à l'établissement d'une relation basée sur le principe de l'égalité des partenaires, il demeure que la place des Cris dans la gouvernance de leurs terres ancestrales devrait s'en trouver améliorée par rapport à la situation qui prévalait jusqu'ici. Au final, la volonté des autorités politiques cries et québécoises de dépasser le cadre conflictuel et de tabler sur la coopération intergouvernementale et le dialogue interculturel doit ici être saluée.

BIBLIOGRAPHIE

Ouvrages cités

Lijphart, Arend. 1999. *Patterns of democracy: government forms and performance in thirty-six countries*. New Haven: Yale University Press.

Morin, Richard. 2006. *Régionalisation au Québec: les mécanismes de développement et de gestion des territoires régionaux et locaux 1960-2006*. Montréal: Saint-Martin.

Nadasdy, Paul. 2003. *Hunters and bureaucrats: power, knowledge, and aboriginal-state relations in the southwest Yukon*. Vancouver: UBC Press.

Proulx, Marc-Urbain. 2002. *L'économie des territoires au Québec: Aménagement, gestion, développement*. Sainte-Foy: Presses de l'Université du Québec.

Rodon, Thierry. 2003. *En partenariat avec l'État: les expériences de cogestion des autochtones du Canada*. Québec: Presses de l'Université Laval.

Articles cités

Christensen, Kyle, et Donley T. Studlar. 2006. « Is Canada a Westminster or Consensus Democracy? A Brief Analysis. » *Political Science and Politics* 39(4): 837-41.

Cooley, Laurence, et Michelle Pace. 2012. « Consociation in a Constant State of Contingency? The case of the Palestinian Territory. » *Third World Quaterly* 33(3): 541-58.

Feit, Harvey. 2010. « Neoliberal Governance and James Bay Cree Governance : Negociated Agreements, Oppositional Struggles, and Co-Governance. » Dans Indigenous Peoples and Autonomy : Insights for a Global Age, dir. Mario Blaser. Vancouver, UBC Press.

Garry, John. 2009. « Consociationalism and its critics: Evidence from the historic Northern Ireland Assembly election 2007. » *Electoral Studies* 28: 458-66.

Glenn, Jane Matthews. 1986. « La décentralisation de l'aménagement du territoire: mythe ou réalité? » *Les Cahiers de droit* 27: 355-70.

Grammond, Sébastien. 2009. « La gouvernance territoriale au Québec entre régionalisation et participation des peuples autochtones. » *Canadian Journal of Political Science* 42: 939-56.

Iyer, Venkat. 2007. « Enforced consociationalism and deeply divided societies: some reflections on the recent developments in Fiji. » *International Journal of Law in Context* 3(2): 127-53.

Lemarchand, René. 2006. « Consociationalism and power sharing in Africa: Rwanda, Burundi, and the Democratic Republic of the Congo. » *African Affairs* 106(422): 1-20.

Lijphart, Arend. 1969. « Consociational democracy. » *World Politics* 21(2): 207-25.

Lijphart, Arend. 1985. « Non-Majoritarian Democracy: A Comparison of Federal and Consociational Theories. » *Publius* 15(2): 3-15.

Otis, Ghislain, et Geneviève Motard. 2009. « De Westphalie à Waswanipi: la personnalité des lois dans la nouvelle gouvernance crie. » *Les Cahiers de droit* 50: 121-52.

Ritter, Archibald R.M. 2001. « Canada: From Fly-in, Fly-Out to Mining Metropolis. » Dans *Large Mines and the Community: Socioeconomic and Environmental Effects in Latin America, Canada and Spain*, dir. Gary McMahon et Felix Remy. Ottawa: IDRC Books.

Saganash, Diom Roméo. 1993. « Le droit à l'autodétermination des peuples autochtones. » *Revue générale de droit* 24: 85-91.

Scott, Colin H., et Jeremy Webber. 2001. « Conflict Between Cree Hunting and Sport Hunting: Co-Management Decision Making at James Bay » Dans *Aboriginal Autonomy and Development in Northern Québec-Labrador*, dir. Colin H. Scott. Vancouver: UBC Press.

Seaver, Brenda M. 2000. « The Regional Sources of Power-Sharing Failure: The Case of Lebanon. » *Political Science Quarterly* 115(2): 247-71.

Sinardet, Dave. 2011. « Le fédéralisme consociatif belge: vecteur d'instabilité? » *Pouvoirs* 136: 21-35.

Spears, Ian S. 2002. « Africa: The limits of Power-Sharing. » *Journal of Democracy* 13(3): 123-36.

Storey, Keith. 2010. « Fly-in/Fly-out: Implications for Community Sustainability. » *Sustainability* 2: 1161-81.

Younis, Nussaibah. 2011. « Set up to fail: consociational political structures in post-war Irak, 2003-2010. » *Contempory Arab Affairs* 4(1): 1-18.

Législation citée

Loi constitutionnelle de 1867, 30 & 31 Victoria c 3 (R.U.)

Loi constitutionnelle de 1982, constituant l'annexe B de la *Loi sur le Canada*, 1982 c 11 (R.U.).

Loi sur la Commission municipale, LRQ c C-35.

Loi sur le Conseil régional de zone de la Baie James, LRQ c C-59.1.

Loi sur le développement de la Baie James et d'autres dispositions législatives, LQ 2001 c 61.

Loi sur le développement et l'organisation municipale de la région de la Baie James, LRQ c D-8.2.

Loi sur le ministère des Affaires municipales, des Régions et de l'Occupation du territoire, LRQ c M-22.1.

Loi sur le régime des terres dans les territoires de la Baie-James et du Nouveau-Québec, LRQ c R-13.1.

Loi sur les cités et villes, LRQ c C-19.

Loi sur les compétences municipales, LRQ c C-47.1.

Loi sur les Cris et les Naskapis du Québec, SC 1984 c 18.

Loi sur les villages cris et le village naskapi, LRQ c V-5.1.

Jurisprudence citée

Bande de Mistissini c. *Iserhoff et Conishish-Coon* (14 août 1996), Abitibi 635-72-000006-959, 635-72-000029-951 (CQ) (j. Boisvert).

Beckman c. *Premières nations Little Salmon/Carmack*, [2010] 3 RCS 103.

Delgamuukw c. *Colombie-Britannique*, [1997] 3 RCS 1010.

Grand Chef Matthew Coon Come et al. c. *Hydro-Québec, le Procureur général du Québec et le Procureur général du Canada*, C.S.M. 500-05-027984-960.

Grand Chef Matthew Coon Come et al. c. *Hydro-Québec, le Procureur général du Québec et le Procureur général du Canada*, C.S.M. 500-05-004330-906.

Eastmain c. *Gilpin*, [1987] RJQ 1637.

Nation Haïda c. *Colombie-Britannique (Ministre des Forêts)*, [2004] 3 RCS 511.

Première nation crie Mikisew c. *Canada (Ministre du Patrimoine canadien)*, [2005] 3 RCS 388.

Première nation Tlingit de Taku River c. *Colombie-Britannique (Directeur d'évaluation de projet)*, [2004] 3 RCS 550.

R. c. *Badger*, [1996] 1 RCS 771.

R. c. *Pamajewon*, [1996] 2 RCS 821.

R. c. *Sparrow*, [1990] 1 RCS 1075.

Rio Tinto Alcan Inc. c. *Conseil tribal Carrier Sekani*, [2010] 2 RCS 650.

Ententes citées

Accord-cadre entre les Cris d'Eeyou Istchee et le Gouvernement du Québec sur la Gouvernance dans le territoire d'Eeyou Istchee Baie James. 27 mai 2011.

Accord définitif de la Première nation Tsawwassen. 6 décembre 2007.

Accord définitif Nisga'a. 27 avril 1999.

Accord sur des revendications territoriales entre les Inuit du Labrador et Sa Majesté La Reine du Chef de Terre-Neuve-et-Labrador et Sa Majesté La Reine du Chef du Canada. 22 janvier 2005.

Accord sur l'autonomie gouvernementale de la Première nation de Westbank entre Sa Majesté la Reine du chef du Canada et la Première nation de Westbank. 3 octobre 2003.

Accord sur les revendications territoriales et l'autonomie gouvernementale entre le peuple Tlicho et le Gouvernement des Territoires du Nord-Ouest et le Gouvernement du Canada. 25 août 2003.

Convention de la Baie-James et du Nord québécois et conventions complémentaires, Québec, Publications du Québec, 2012.

Entente concernant une nouvelle relation entre le Gouvernement du Québec et les Cris du Québec. 7 février 2002.

Entente concernant une nouvelle relation entre le gouvernement du Canada et les Cris d'Eeyou Istchee. 21 février 2008.

Entente sur la gouvernance dans le territoire d'Eeyou Istchee Baie-James entre les Cris d'Eeyou Istchee et le Gouvernement du Québec. 24 juillet 2012.

Entente sur certaines questions liées à l'entente sur la gouvernance dans le territoire d'Eeyou Istchee Baie-James entre les Cris d'Eeyou Istchee et le Gouvernement du Québec. 24 juillet 2012.

Documents gouvernementaux cités

Gouvernement du Canada. 1995. *Guide de la politique fédérale sur l'autonomie gouvernementale: l'approche du gouvernement du Canada concernant la mise en œuvre du droit inhérent des peuples autochtones à l'autonomie gouvernementale et la négociation de cette autonomie*, Ottawa.

Gouvernement du Québec. Ministère du Conseil exécutif. 2004. *Devenir maître de son développement. La force des régions. Phase 1 – Une nouvelle gouvernance régionale*. Québec: Secrétariat à la communication gouvernementale.

Gouvernement du Québec. Conseil exécutif. 2009. *Rapport annuel de gestion 2008-2009*. Québec.

Gouvernement du Québec. 2012. *Communiqué de presse du 24 juillet*. Québec.

Documents des Nations Unies cités

Déclaration des Nations Unies sur les droits des peuples autochtones. 2007. Rés. 295. Doc. off. A.G. N.U. 61ᵉ sess. Doc. N.U. A/RES/61/295.

Communications citées

Grammond, Sébastien. 2008. « La gouvernance territoriale et l'obligation constitutionnelle de consulter et d'accommoder les peuples autochtones » présenté dans le cadre du colloque de l'ASRDLF, Rimouski. [En ligne] [http://asrdlf2008.uqar.qc.ca/Papiers%20en%20ligne/GRAMMOND-S.pdf] (21 août 2012).

Le comité organisateur. 2012. « Synthèse des discussions et des échanges tenus lors de la table ronde » dans le cadre du Colloque « Routes et faune terrestre: de la science aux solutions », Québec. Publié dans *Le Naturaliste canadien* (2012) 136: 107-08.

Murdoch, John Paul. 2008. « The Future of Governance in Eeyou Istchee » présentée au 6e Forum Autochtone sur la gestion des ressources naturelles et du territoire, Montréal.

Rapports cités

Juge Réjean F. Paul. « Rapport » 4 novembre 2008.

Juge Réjean F. Paul. « Rapport » 4 mars 2010.

The Grand Council of the Crees. « Cree vision of Plan Nord » février 2011. [En ligne] [http://www.gcc.ca/pdf/Cree-Vision-of-Plan-Nord.pdf] (10 août 2012).

Lettre citée

Iserhoff, Ashley. Lettre du président du Comité pour l'environnement de la Baie-James adressée à Monsieur Robert Sauvé, alors sous-ministre du Ministère des Ressources naturelles et de la Faune. 28 avril 2011.

Mémoire de maîtrise cité

Frouin, Hermann. 2011. « Influence des corridors routiers et des coupes sur les déplacements hivernaux de la Martre d'Amérique en forêt boréale aménagée. » Mémoire de maîtrise. Université Laval.

9

THE POLITICS OF REGIONS AND RESOURCES IN AUSTRALIA

Douglas Brown

INTRODUCTION

Canada and Australia are both resource-producing giants. The long resource boom has brought enormous benefits but also some costs to our two countries. This volume addresses the politics of regions and resources in the large and diverse Canadian federation. Australia is a natural point of comparison as we examine our own issues, problems, and potential solutions. Like Canada, Australia is a large, territorially diverse country with a federal constitution and many similar political institutions. It, too, is a multicultural settler society with a coexisting indigenous population, a wealthy, advanced industrial economy well integrated globally, and an Anglo-American business culture. Unlike Canada, Australia is an island continent, relatively isolated – it is not attached geographically to a much more populous continental neighbour, and thus is not as dependent on a single major trading partner as Canada is on the United States. While resource and energy production and markets in Australia do differ from Canada's,[1] the issues surrounding terms of trade and economic adjustment, environment (including greenhouse gas emissions), interregional income, and labour force balance, among others, are all very similar to the issues that have arisen in Canada in the past decade.

This chapter examines recent resource and energy politics in Australia and how they are affected by or in turn influence federalism and intergovernmental relations, including federal values and the overall ability of the system to manage conflict and change. The next two sections explore the broader context: first, the

I wish to thank Andrew Banfield, Robert Milliken, and the editors of this volume for their helpful comments.

[1] For further details see Grant (2013) and Blackwell (2013).

political economy of regions and resources in Australia; and second, Australian federal values, institutions, and practices. Then we turn to three brief cases of policy issues in Australia that have relevance for Canada: the National Electricity Market, the regulation of greenhouse gas emissions, and recent federal-state issues over resource revenues and the fiscal equalization system. The chapter concludes with lessons for Canada.

THE POLITICAL ECONOMY OF REGIONS AND RESOURCES

With 22.6 million people in a territory of 7.7 million square kilometres, Australia is not densely populated, but it does have concentrated patterns of population distribution (see Figure 1 map). Essentially most of the population and all of the bigger cities are on or near the coasts, reflecting initial settlement patterns and colonial port development. The most populous states of New South Wales and Victoria, including the two largest cities of Sydney and Melbourne, are in the southeast, as is the Australian Capital Territory where Canberra is situated. The agriculture and agri-foods industry is a major industry in all states; non-agricultural manufacturing tends to be concentrated in Victoria, South Australia, and New South Wales. The mining sector, while present in all states and territories, is most prominent in Western Australia and Queensland. Commodity exports have been important to the Australian economy since at least the 1850s, with wool, mutton, beef, forest products, and gold dominating early on, and other minerals later in the twentieth century. The modern service economy now exists everywhere, especially in the urban areas where over 90 percent of people live (depending on the definition of "urban"). Up to the mid-twentieth century, the traditional ethos in Australia was one of the country living "off the sheep's back." Spatially and geopolitically the economy grew through and out of the main port cities, which in all important cases are also the state capitals. Each state had its tentacles into the agriculture and other resources of its hinterlands. In fact, almost all of the key economic infrastructure originated in each individual state as well, much of it state-owned and run, such as railways, port facilities, and electricity generation and distribution. Only with the nationally coordinated microeconomic reforms of the 1990s did governments begin a long-term effort at infrastructure integration (Brown 2002; Quiggan 1996).

Regional diversity and disparity is clearly present in Australia even if it is not quite the issue it presents in Canada. The structure of the state economies is not as dissimilar; there is not the same history of regional alienation and quasi-colonial domination of one region over another, and national policy did not generate as much regional conflict as in Canada. Until recently, the terms of trade of Australia's resource wealth tended to pull or push most parts of the country in the same direction (Courchene 1996). Recently, population has been moving north and west – to

Figure 1: Political Map of Australia, Population and GDP Ratio by State and Territory, 2012

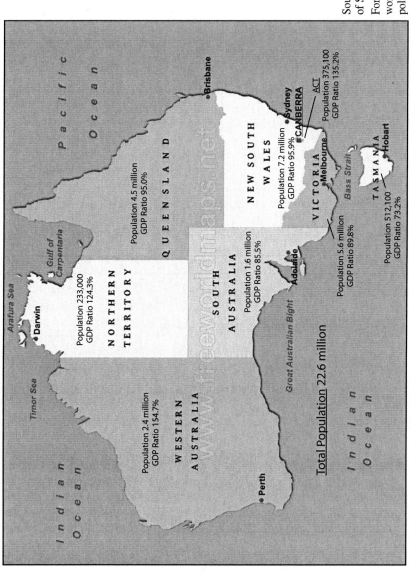

Population 233,000
GDP Ratio 124.3%

Population 2.4 million
GDP Ratio 154.7%

Population 4.5 million
GDP Ratio 95.0%

Population 1.6 million
GDP Ratio 85.5%

Population 7.2 million
GDP Ratio 95.9%

Population 375,100
GDP Ratio 135.2%

Population 5.6 million
GDP Ratio 89.8%

Population 512,100
GDP Ratio 73.2%

Total Population 22.6 million

Source: Australian Bureau of Statistics (2012).
For map: http://www.free worldmaps.net/australia/ political/html.

"sunshine states" – and to the growing resource development areas of Queensland and Western Australia. Figure 1 shows considerable diversity in population size among the states and territories, but also in GDP/capita. Four states or territories stand out.[2] The higher GDP/capita in the Australian Capital Territory and Northern Territory is explained in the first case by a small population largely employed in high-income public services, and in the second by a small population in the presence of some significant resources. Tasmania is a small, poorer province without significant resources, whereas Western Australia has a medium-sized state population with an enormous resource economy (home of the biggest iron ore deposits and the most natural gas). The fiscal capacity of the states in 2013 ranges from Western Australia with 181 percent of the national average to Tasmania with 63 percent (CGC 2013). Unemployment rates in July 2012 ranged from 3.8 percent in Western Australia to 7.3 in Tasmania, with most states clustered near the national average of 5.1 (ABS 2012).

Moreover, the term "region" has a somewhat different meaning in Australia, where "regional Australia" is taken to mean everything that is non-urban or non-suburban: all the small towns, rural and remote communities and outback, cutting across all the mainland state and territorial boundaries. Otherwise, Australians refer to issues as being interstate, sometimes west-east (usually referring to Western Australia versus the rest), sometimes north-south (Queensland and the Northern Territory versus the southeast). Apart from the important urban/rural divide, the most clearly separate "regions" in a geographical and political culture sense are the states of Western Australia, Queensland, and Tasmania, and the Northern Territory. Tasmania is an island, obviously separated from the mainland states, while the three units of Western Australia, Queensland, and the Northern Territory are more sparsely populated and much farther than the other states from the concentrated population of the southeast and from the federal capital in Canberra.

Nonetheless, due in part to the nature of Australia's political history (a strong working class and social democratic influence), federal design (centralized, as discussed below), and more homogenous society, state-based regional disparities as such have been muted. In practical terms the evidence for this is in centralized wages and other working conditions applicable throughout the country since before the First World War (being dismantled only in the past two decades), in the effects of a comprehensive fiscal equalization system in place since the 1930s, in the wider scope (than Canada) of social benefits provided directly by the federal government, and in the more limited scope (than Canada) of regulatory and fiscal intervention by state governments as opposed to the federal government, although as noted states have played a major role in infrastructure (Courchene 1996).

---

[2] The territories have a basis in federal legislation and have most, but not all, state powers. Constitutionally they are not constituent units as such and do not have a role in constitutional amendment.

The Australian economy has not been in recession since 1991, and in particular weathered the Asian financial crisis of 1997 and the global recession and financial crisis of 2008–09 arguably better than any other major economy. The "long boom" in Australia has been driven by enormously beneficial terms of trade for resources and agriculture,[3] in turn clearly linked to the growth in demand for Australian output in iron ore, coal, and natural gas in the global economy in general and in China and southeast Asia in particular. Moreover, Australia managed to avoid a recession in 2008–09 because its financial sector had little exposure to the corrupted mortgage loan markets in the United States, and was not especially dependent on US or European domestic demand for its exports. From 2000 to 2010, the mining sector doubled its share of GDP, now at 10 percent, the same as manufacturing. Planned investment in the mining sector in 2013 alone has been estimated at 9 percent of GDP (*Economist* 2012f).

The problems associated with resource-led economic growth are very familiar to Canadians, with some key differences. These problems can be clustered in a number of categories including macroeconomic effects on the economy as a whole, micro-economic effects on certain sectors, environmental issues, and fiscal policy issues. In all of these there is a clear regional and spatial component. To begin, Australians do worry about the so-called Dutch disease; while they do not have a "petro-dollar" as such, the terms of trade of the key commodity prices have kept the Australian dollar above the US dollar for several years. The manufacturing and some service sectors blame the resource economy for pricing exports too dearly and raising the costs of labour. Whether this can all be attributed to the resource boom can be doubted, but 100,000 jobs have been lost in the manufacturing sector, mostly in Victoria and South Australia (*Economist* 2012a, 2012e; Victoria 2011; compare Coulombe 2013). Otherwise, commentators worry about Australia being "China's quarry," about resource dependence, and about excessive foreign investment in the mining sector (*Economist* 2012f). Other issues arise from the size and scale of the resource developments themselves, which drive up labour and housing prices in the regions where they take place – contributing to two-speed economies, even within a single state. State governments worry about the provision of service and infrastructure at inflated costs, and the intrastate political tensions of dealing with overheated local economies.[4]

Environmental issues are often site-specific dealing with common mineral development concerns, which have on occasion become national issues, but by far the biggest set of concerns has been related to greenhouse gas emissions (GGEs). Here the regional issues are important, even though – and this is quite significant in

---

[3] The ratio of export to import prices is "the most favourable in 140 years." *The Economist* (2012c).

[4] Interviews with senior officials in the Department of Premier and Cabinet, and the Department of the Treasury, Western Australia government, Perth, October 25–26, 2013; Western Australia (2011).

comparison to Canada – the regional incidence of GGEs is not nearly as concentrated (see Table 1). Australia, like Canada and the United States, is a carbon-intensive economy with relatively cheap gasoline prices and dispersed urban populations, is heavily dependent on automobiles, and has had strong economic and population growth. Australia has the highest level of carbon emissions per capita among the OECD countries. It is a major domestic consumer of coal for electricity in addition to exporting coal for electricity production elsewhere. The three largest coal-producing states are Victoria (brown coal mainly for domestic energy consumption), New South Wales, and Queensland (black coal, with most production exported). On the whole, the resource production sectors including agriculture and mining are major GGE emitters (Crowley 2010; Garnaut 2008). As will be discussed more fully below in relation to the federal carbon price "tax," the issue of regulation of GGEs is not just an issue of regions and resources, but the latter are important factors.

Table 1: Greenhouse Gas Emissions by State and Territory, 2010
(Million tonnes CO2)

	Volume	*Per Capita*
New South Wales	157.4	21.6
Queensland	157.3	34.7
Victoria	117.9	21.1
Western Australia	74.3	30.8
South Australia	29.3	17.8
Northern Territory	14.7	63.0
Tasmania	7.6	14.8
Australian Capital Territory	1.2	3.2
Total Australia	**560.8**	**24.8**

Source: Commonwealth of Australia website, http://www.climatechange.gov.au/ publications/, accessed August 13, 2013.

Finally, there is the issue of fiscal resources. Natural resources onshore are owned by the states,[5] and they have the right to levy royalties. Otherwise, fiscal federalism is considerably centralized in Australia, such that the federal government is in a position to benefit directly from natural resource growth and development. The enormous profits (including windfall profits) of the mining companies in particular have become a magnet for the federal government. The controversy over a short-

---

[5] Offshore resources are owned by the federal government, although in the case of one natural gas development the federal government agreed to share royalties with Western Australia; see Anderson (2012).

lived federal minerals tax is discussed below. Federal taxes are by far the most significant taxes and are redistributed extensively. Moreover, in the long-standing practice of fiscal transfers, richer states (including newly richer states) contribute through equalization to the redistribution to poorer states (it is a net scheme as applied to GST revenues), a situation that is intensified for states such as Western Australia in the current resource boom. As such then, issues over resource income – including the carbon price "tax" – tend to pit state against state and state against federal government in a not quite zero-sum game.

Federal Values, Institutions, and Practices

Australia was created as a union of six former British colonies, and not as a union of peoples, nations, or linguistic groups. Australia's rights are more conventional (in the legal sense) and more implied. There is no entrenched bill of rights, so parliamentary supremacy still prevails, divided between the federal government and the states. On the other hand, as Brian Galligan (1995) argues, Australia's constitution is republican in that sovereignty resides more obviously in the people. Even though it retains the British monarch as the head of state, the 1901 constitution was framed by directly elected delegates and ratified by popular referendum before being formally enacted by Britain. Australia's federal constitution, like Canada's, was grafted onto Westminster-style parliamentary government. The basics of responsible government and evolution of party discipline and prime ministerial power are very similar. A key difference is Australia's elected Senate in the federal Parliament, from which the governing prime minister draws cabinet ministers as well as from the lower house. The Senate has equal representation per state, but party discipline is only slightly moderated by regional interests. A more significant feature is the Senate's electoral system by statewide franchise, which enables independents and smaller parties to be represented, sufficient in recent decades to provide the governing party with only a minority of seats (Brown, Bakvis, and Baier 2011). Most major legislative policy of the Commonwealth (federal) Parliament is in effect the result of bargaining between the two houses. In sum, applying Smiley and Watts's (1985) "intrastate federalism" concept, the Australian Senate performs effectively if not perfectly in representing the interests of the constituent units.

On the distribution of powers, the Australian constitutional founders chose the American federal model of concurrent powers to be shared between the federal and state legislatures, with the states holding the reserve. Federal laws are paramount in the event of conflict, which has had a significantly centralizing effect over time. Moreover, the High Court in Australia has never felt as bound as its Canadian counterpart to emphasize federal balance when faced with state legislation, but rather sees its role as providing for an increasing degree of national legal integration, if not centralization (Coper 1988). This is not to exaggerate the point: states still have substantial legal and administrative jurisdiction as subnational entities. And in at least one important area, criminal law, the states have primary jurisdiction. Overall, however, the Australian distribution of legislative jurisdiction is relatively centralized.

One consequence of centralized legal power is centralized fiscal power (Saunders 2011, 237-43; Ward and Stewart 2010, 137-41). The rather narrow initial allocation of states' taxing powers has been narrowed even further by High Court decisions favouring the federal government, in particular in the 1940s and 1950s to confirm federal control over the income tax field, and in later cases that keep the states out of the consumption tax field. As a result, the central government now levies five out of every six dollars in taxes in Australia, coupled with an explicit and broadly interpreted spending power in which state jurisdiction seems to matter very little. The states are left with gaming revenues, resource rents, and a variety of small tax sources as well as property tax powers delegated to local government. A severe vertical fiscal imbalance is alleviated by substantial intergovernmental transfers, both conditional (special purpose payments, national partnership payments, among others) and unconditional (almost all in the form of the GST distribution noted above), totalling $44.1 billion and $51.2 billion respectively in 2013–14 (Commonwealth 2013). The average state reliance on federal transfers is around 45 percent (see Figure 2).

A final federal feature is the system of intergovernmental relations. Australia being a parliamentary federation also exhibits "executive federalism" whereby relations are concentrated in the executive branch. This means they are monopolized by first ministers, other cabinet ministers, and their senior officials, to the general exclusion of legislatures (Watts 1989, 2008). Formal and informal relations among governments became much more intense with the increased role of the state after the Second World War, particularly with the build-up of social programs financed by intergovernmental grants. As de facto government roles became less divisible and more interdependent, the significance of intergovernmental relations to the overall policy-making process and to the political system as a whole increased. Trends in the development of executive federalism in Australia were very similar to those in Canada until the 1990s. At that point, as part of a shared agenda of microeconomic and fiscal reform tied into Australia's competitiveness strategy, Australian governments decided that they needed to significantly upgrade their intergovernmental mechanisms to achieve a more comprehensive and coordinated reform program (Brown 2002; Painter 1998). While the initial reform episode has long passed, an institutional legacy of upgraded intergovernmental capacity continues. What one can call the Australian model of intergovernmental co-decision consists of the following elements:

- the formal establishment of a Council of Australian Governments (COAG)[6] to meet at least annually, which in practice often meets more frequently;

---

[6] COAG membership is restricted to the federal prime minister, the six state premiers, the two territorial chief ministers, and the president of the Australian Local Government Association.

- a rationalized and streamlined set of standing and ad hoc Ministerial Councils (MCs), under the scrutiny of COAG, if not always reporting directly to it;
- MCs that can take binding decisions, backed up by uniform federal and state and territorial legislation;
- voting rules in these MCs that allow the councils to take decisions by majority or qualified majority vote;
- several new joint "national" agencies in fields such as energy, environment, food standards, road transport, training, and competition policy; coordination through non-centralized devices such as mutual recognition of standards, and "negative integration" through such policies as national competition; and
- from 2007 to 2014, the COAG Reform Council, an independent body, provided assessments of whether governments were meeting commitments made in intergovernmental undertakings.

Figure 2: Aspects of Fiscal Arrangements – Australian Federal System

Major Federal Tax Assignments (100% of field unless otherwise specified):
- Personal income tax
- Corporate tax
- Goods and services (sales) tax
- Excise tax
- Payroll taxes (25%)

Major State Tax Assignments (100% of field unless otherwise specified):
- Payroll taxes (75%)
- Land tax
- Financial services tax
- Gambling tax
- Motor vehicle registration
- Mining revenue
- Property (100% levied by local government)

General Government Revenues and Expenditures as Percentage of GDP

	Federal Revenues	Federal Expenditures	State Revenues	State Expenditures
2000–01	26.3	25.5	16.3	15.9
2005–06	26.3	24.4	16.3	15.3
2010–11	22.1	25.4	16.2	15.7
2013–14	24.3	24.9	15.5	15.4

Sources:
On tax assignments: Morris (2007), Table 4.
On revenues and expenditures: Commonwealth (2013), Table C4.

Contrasted with the status and practice of Canadian intergovernmental relations, these innovations go considerably further to ensure the capacity to reach substantive, binding, joint decisions. The chief (and significant) caveat is that the new machinery in Australia requires political will, particularly by the federal prime minister, to make it work. That political will comes and goes, but in 20 years all federal governments have made at least minimal use of the new system; in fact, the governments as a whole have achieved a very impressive record of agreement on a wide range of issues.

These institutional features of federalism – an effective regional voice in the federal Parliament, a concurrent distribution of powers, centralized fiscal arrangements, and muscular intergovernmental processes – are underpinned by specific federal values. In Australia there is obviously less of a "federal society,"[7] despite a resurgence of Aboriginal identity and multicultural diversity. As noted, regionalism is less evident, and except for a fleeting separatist movement in Western Australia in the 1930s, no aspects of regionalism can be confused with nationalism. Moreover, there is simply less tolerance for state diversity in policy and political outcomes than there would be in other federal systems, such as Canada. Instead there is strong preference in both public opinion and among elites for national harmonization, uniformity, and equity.

A century of national integration and nation building reinforces Australia's more centralized federal values. The nation-building process is fostered by a highly integrated federal and state party structure and by the federal Parliament's clear notion of national leadership. Unlike Canada where the notion of a national strategy or policy can be highly problematic, there are little or no constraints on the articulation and development of national strategies within federal (i.e., central) institutions, and much more propensity toward cooperation and collaboration with the states and territories in the application and implementation of national policy. Australia's federal values, institutions, and practices are well illustrated in the recent cases of resource and energy policy initiatives presented next.

RECENT NATIONAL POLICY INITIATIVES

For the purposes of this chapter, I have selected three sets of cases or initiatives dealing with current or ongoing intergovernmental relations on energy and resource issues. At least two of them (greenhouse gas emissions and the mineral tax) are very

---

[7] A federal society refers to the extent to which the federal political community is one where power is shared not only among territorial units but also among linguistic, cultural, ethnic, religious, or national communities. That power sharing can be implicit or explicitly organized into federal institutions. Different federal systems encompass differing degrees of federal society. For discussion see Watts (2008, 19-21), citing in particular Livingston (1956) and Cairns (1977).

significant current political issues, but they have also been chosen to illustrate how the Australian system delivers results relevant to Canadian problems.[8]

National Electricity Market

Electricity market reform was one of the many targets of microeconomic reform in Australia from the mid-1980s. Key problems identified were the lack of competition among electricity providers, the inefficiencies of state-based and often state-owned electricity systems, and the absence of a national electricity grid. The federal Labor government under Robert Hawke began a wide-ranging initiative in 1991 to engage the states and territories in a national economic reform strategy, agreeing as a part of that process in July 1991 to develop a National Grid Management Council. By 1998 that council had established a national electricity market, initially between the two most populous states of New South Wales and Victoria, and since extended to all eastern jurisdictions (i.e., Queensland, South Australia, Tasmania, and the Australian Capital Territory; the main markets in the Northern Territory and Western Australia are likely too far away to ever be feasibly part of the national market). The market took much political effort to achieve given differing ideological and strategic agendas among the states (some retaining state ownership, some going to privatization and market competition reform and so on). The National Electricity Market (NEM) now covers about 8 million customers and $11 billion in wholesale transactions: all major public and private electricity providers in the participating jurisdictions are included.

The institutions that govern the NEM are complex, even Byzantine to the outside observer. Explicitly mandated by COAG and by the relevant federal-state-territorial Ministerial Council (i.e., the Standing Council on Energy and Resources), the arrangements were established in a 1996 template act of the South Australian Parliament and applied in mirror legislation in all of the other participating states and territories, as well as the federal government. There are two key regulatory bodies, one an independent, arm's-length commission within the federal public service establishment, the Australian Energy Market Commission (AEMC),[9] which also deals with a national gas market. It both advises the intergovernmental ministerial council on policy matters, and plans and initiates the NEM "electricity rules" that establish the parameters of the spot market and other aspects. It supervises the actual "Operator" of the grid. There is also an Australian Energy Regulator that enforces the rules. The Ministerial Council recently handed the AEMC an additional mandate to enact customer protection provisions across the

---

[8] Other issues that are worth pursuing, but not here due to space constraints, include water resources, labour standards and mobility in the resource economy, Aboriginal land claims and affirmative action employment, and foreign investment in the resources sector.

[9] See www.aemc.gov.au, accessed June 4, 2013.

grid. The goal is for all participating states and territories to transfer authority to the federal government to establish a single, national set of rules governing retail customer-utility relationships.[10]

Climate Change and Greenhouse Gas Emission Regulation

Australia participated fully in the initial negotiations for the United Nations Framework Convention on Climate Change at Rio de Janeiro in 1992. The federal government signed on to the Convention, and participated fully in the diplomacy leading to the Kyoto Protocol reached in 1997. However, led by the conservative coalition under Prime Minister John Howard from 1996 to 2007, the federal government was opposed to the Kyoto Protocol. Only after an election in 2007 brought the Australian Labor Party to power under Kevin Rudd did Australia ratify the agreement. Nonetheless, during the Howard years, and under the auspices of COAG and the relevant Ministerial Council, the federal and state and territorial governments reached early agreement on a number of key issues dealing with greenhouse gas emissions: undertaking a national carbon storage initiative, establishing national emission reporting, endorsing various coordinated efforts at energy efficiency promotion, and doing some initial work in establishing national emission targets (Gordon and MacDonald 2011). COAG could not get very far on the key regulatory issue of carbon pricing so long as Howard's coalition government opposed it, but in the meantime the state governments, many of them Labor, took action on their own. Prevented by constitutional law from imposing carbon taxes as such, the states concentrated on energy efficiency programs and, most importantly, renewable energy mandates for their publicly owned or regulated electricity utilities. Intriguingly, the state premiers in 2006 banded together to call for Kyoto ratification and a national plan for GGE reductions, and commissioned a major independent report (Garnaut 2008) detailing the need for Australian leadership on the issue. These state governments were responding to the significantly higher public profile of climate change issues in Australia by the mid-point of the 2000s, which the Howard government found more difficult to do. The states' actions were partisan, but nonetheless succeeded in laying groundwork for the national legislative approaches that followed with the ALP (Australian Labor Party) federal governments led by Kevin Rudd and Julia Gillard.

Prime Minister Kevin Rudd, whose ALP party defeated Howard in 2007, attempted to pass a comprehensive cap and trade regime in 2008–10 but was defeated in the Senate of the federal Parliament, where the bill was condemned by the coalition of conservatives as too much and by the Greens as too little (of course the ALP did not have a majority in the Senate). Following the 2010 election, ALP

---

[10] For details see www.scer.gov.au, accessed June 4, 2013.

prime minister Julia Gillard succeeded in passing through the federal Parliament a revised national carbon pricing regime with the support of the Green party. Their scheme, which they called Clean Energy Future (CEF), came into effect in July 2012.[11] The CEF established an initial carbon price fixed at $23/tonne to be paid by the 300 largest emitters. Emission trading was set to begin in 2015, after which the carbon price would be determined by that market (Crowley 2010; Harrison 2011; Macintosh, Wilkinson, and Denniss 2010).

There is a widespread recognition in Australia, including among the state governments, that once the federal Parliament decides to act on greenhouse gas emission regulation, its actions would trump anything the states would do because of its paramount jurisdiction over the environment, its sweeping treaty execution powers, and the long history of states giving way to national strategies when consensus is formed in their favour. The notion that national legislation on GGEs could intrude too far on state autonomy would occur to very few Australians. Moreover, while Australia exhibits some significant variation in GGEs by state, with major mining states (Queensland and Western Australia) and those with especially high coal-fired electricity production (notably Victoria) standing out, the overall pattern of carbon use is more uniform than in Canada, blunting direct regional conflict (see Table 1). The interests such as the mining and resources sectors, agriculture, and other business that have consistently opposed national regulation have found their champions in the conservative parties, and through lobbying have been able to wrest some concessions in the Gillard government's CEF, for example the general exemption for agriculture. They were hoping that a conservative coalition government, if elected in the next federal election, would overturn the CEF. Indeed, the carbon tax/price regime is now essentially dead after the election of the Liberal-National coalition in September 2013. The new government promised to scrap the regime in favour of an emissions reduction fund.[12]

In sum, the Australian federal system has delivered on climate change across the board: first, there is a federation-wide understanding of what global warming and climate change mean, including a common understanding of the need to move toward a carbon-reduced economy and society. Second, there has been agreement, not without some contestation, on Australia's contribution to the global problem, although there is a lot of emphasis on a post-Kyoto regime. Third, the federal government, despite being initially opposed to the specifics of Kyoto, did actively plan with the states to determine total GGE incidence and allocation. Rather, the debate over the Clean Energy Future plan was about how quickly and the extent to which certain major sectors in Australia (coal-fired utilities, steel, aluminum,

---

[11] For details see http://www.cleanenergyfuture.gov.au/, accessed September 23, 2012.

[12] For a comparison of all the party positions on climate change issues see "Environment Policy: Where the Parties Stand," http://www.abc.net.au/news/federal-election-2013/, accessed August 28, 2013.

mining, etc.) were to contribute to GGE reductions, and of course about whether a new "tax" was the best approach. Fourth, Australia actually achieved, unlike Canada or the United States, a binding national regime for GGE reduction, even if the new federal government under Prime Minister Tony Abbott has pledged to scrap the policy in favour of a "direct action" emissions reductions fund. In any case, there is no doubt that any federal scheme will be binding on the states.

Resource Revenue Issues

The states and territories own the mineral resources onshore and levy a complex set of royalty and related taxes, usually based on production volumes not value, geared historically to promoting development in the context of struggling markets and prices. In the long resource-commodities boom that started about 2002, only recently faltering (Blackwell 2013), mining company profits have attracted attention, as has the regionalized economic boom over hugely increased mining sector development. Australia's petroleum resources are mostly in the form of natural gas. These resources until recently have been mostly in offshore deposits off the southeast and (especially) northwest coasts. There are currently major new developments for coal seam and shale gas deposits onshore. Historically, public revenues from petroleum resources have been quite small in comparison to those in Canada. Nonetheless, resource revenues are increasingly an important intergovernmental and interregional issue as illustrated by two continuing controversies: the treatment of resource-based economies by the overall fiscal equalization system, and the entry of the federal government into the mining tax field.

A major component of Australia's system of intergovernmental grants is the distribution of the revenues from the federal government's Goods and Services Tax (GST). The arrangement dates from 1999 when the Howard government reached agreement with the states and territories, through the COAG process, to distribute all of the revenue (minus collection costs) from the new GST to the states on an equalized basis (Brown 2002, 218-22).[13] The agreement was a way both to reform tax structure by introducing a new, broad-based national consumption tax at the same time as abolishing many inefficient state taxes, and to reform fiscal relations by providing states with a source of unconditional funding tied to a growing revenue source (a pool that yielded $A51 billion in fiscal year 2013–14). The equalization process involves the determination of relative fiscal capacity for each state and territory by the Commonwealth Grants Commission (employing methodology and principles used for decades) applied to the GST revenue pool. States with

---

[13] For the text of the 1999 agreement, see http://www.coag.gov.au/node/75, accessed June 4, 2013. On current issues about the GST distribution and the equalization process, see Commonwealth (2012).

lower fiscal capacity get revenue higher than the tax receipts actually collected in their state (e.g., in 2013 Tasmania got 158 percent) and states with higher fiscal capacity receive less than the tax receipts from their state (e.g., in 2013 Western Australia got 55 percent; CGC 2013). The relative fiscal capacities have of course been widening during the long boom – mainly due to Western Australia's fiscal capacity galloping away from the other states. This reflects increasing yields from state-based taxes, including resource royalties as well as other tax sources that have risen with increasing economic activity. On the mining royalty effect, however, a Western Australian state official told me that the "effective clawback on GST allocation is huge."[14] Western Australia in particular is seeking some relief in the rate at which its revenue growth related to resource development and production is implicated in the equalization formula. Not only are state officials concerned with the deterioration of Western Australia's revenue position over time, they also see it as a disincentive to modernize mining resource agreements (i.e., to more effectively tax rising mineral values).

Complicating the GST distribution issue is the politics of a new federal minerals tax. In 2010 Labor prime minister Kevin Rudd picked up a recommendation of a recent review of tax policy in the form of a Resource Super Profits Tax, announcing the intention to proceed with little consultation. The new federal tax would be for 40 percent of profits of mining companies covered, but promising to reimburse companies for state royalties paid: "In effect the government was to become an equity partner in resource projects bearing 40 percent of all costs and 40 percent of all economic rents" (Garnett and Lewis 2010, 192). Caught off guard by the announcement, the mining industry and the states launched a withering campaign on the Rudd government, which was already damaged by an overly large agenda and indecision (Aulich and Evans 2010). After Rudd was replaced as leader and prime minister by the federal Labor caucus, Prime Minister Julia Gillard reintroduced a pared-down version called the Mineral Resource Rent Tax (MRRT), levied at 30 percent and on the iron ore and coal sectors only (Australia's two largest exports), and after having negotiated the details with the largest mining firms. The federal government proposed to use the estimated $11 billion in proceeds over three years to fund infrastructure, pensions, and tax cuts (*Economist* 2012b). Legislation for the new tax passed in March 2012, but declining iron ore prices knocked down considerably the optimistic revenue projections, with yields in the 2013 fiscal year expected to be $A200 million compared with the $A3 billion once expected (Ker 2013).

The states, led by the Barnett (Liberal) government of Western Australia, in general opposed the MRRT, not only on the principle that the new tax encroaches at least politically if not constitutionally on one of the few remaining significant own-source revenues of state governments, but also on the fact that it hugely complicates the ability of individual states to get the royalty/tax mix right for

---

[14] Interview with senior Western Australia treasury officials, Perth, October 25, 2011.

optimum development and return for the public owners of the resource and for private investment (*Economist* 2012d; Western Australia 2011.[15] In short, there was potential for a form of tax war as the states increased their royalties, which would reduce federal revenue from the tax.

Both the equalization clawback issue and the conflict over mining resource revenues were taken up recently by an independent panel appointed by the federal government. The panel carefully listened to the views of the states. Its report bluntly stated that

> the impasse between the two levels of government on this [resource revenues] issue is harmful and unsustainable, but […] it won't be fixed by penalizing the States through the GST distribution system. The panel concludes that what is needed is for the States and the Commonwealth to settle a negotiated income. Ideally such an agreed position would enable State royalties to be lower and the revenue from the Commonwealth resource taxes to be greater. (Commonwealth 2012, 4)

This advice is now moot as the new Liberal-National coalition federal government, elected in September 2013, abolished the ALP's mining tax, leaving the states free rein.

CONCLUSION: LESSONS CANADA CAN LEARN

This chapter has deliberately presented issues about regions and resources in Australia within the context of the federal system and other political features of the country, as well as the broader geographic, economic, and social context – all of which differ in important ways from Canada. So at first blush the prospects for Canadians in fact applying Australian models can seem very remote. Still, the value of comparative analysis is often as much in what one learns about one's own country as the other. So, without necessarily treating Australia as a paragon, there are several points that should be taken away from the brief discussion in this chapter. These can be organized first in terms of the capacity for national policy-making and how Canadians might be able to do better to achieve effective results, and second in terms of comparing actual policy outcomes on key resource issues, and where we might want to emulate or avoid the Australian example.

The first systemic finding about Australia is the obvious point that it is a much more centralized federation than Canada. This is due, as we have seen, to both institutional and societal factors. On the social side, there is no linguistic divide, and despite elements of Aboriginal resurgence and multicultural diversity, the prevailing ethos in the federation is not so much one of bringing together a diverse society as it is one of creating a single nation of pre-existing colonies spanning immense

---

[15] Interview with senior Western Australia treasury officials, Perth, October 25, 2001.

geographical and territorial diversity. Also, geographically it matters that Australia is an island continent that for much of its history has felt physically isolated from its cultural heritage – so different from the shared continental experience Canada has had with the United States. So, one need not overplay social homogeneity in Australia to acknowledge that federal politics there is about nation-building. It has been in Canada too, but there are limits to what we want to achieve because our federal ethos is also about nation-preserving with respect to francophones and Quebec and, more recently, Aboriginal peoples.

On the institutional side, several aspects of the evolved Australian political structure reinforce and reflect the greater social unity. These include the adoption of a concurrent division of powers among the federal and state legislatures; an elected Senate that shares power with the federal House of Representatives;[16] an integrated federal and state party structure; centralized fiscal federalism; and, most recently, the adoption of more formalized intergovernmental machinery capable of co-decision. The upshot is a political community more capable than Canada's of achieving nationally cohesive and coordinated results and, as noted, one that on almost every measure can be considered as more centralized than Canada's. There remain major obstacles to achieving political consensus in Australia, but they tend to be ideological rather than regional or national. (And the reinforcing of ideological with regional cleavages in Canada does not occur to anywhere near the same degree in Australia.) Some may wonder if Australians still want a federal system at all. However, the relative centralization belies vital and robust state political communities, and the Australian people have voted down a series of constitutional amendments designed to diminish the states (see Fenna 2007; Galligan 1995; Saunders 2011).

Taking for granted that the federal society in Canada is fundamentally different from Australia, close observation of Australian institutions might lead nonetheless to consideration of institutional reform in Canada. The two obvious candidates are the Senate and intergovernmental relations. Elsewhere I have argued that an elected Senate on the Australian model brought to Canada would not likely operate quite the same way as it does there (Brown, Bakvis, and Baier 2011). It seems very unlikely that it would be dominated by the same parties as dominate the lower house; that is, the parties in the Canadian Senate would likely be more regionally based, and therefore the Senate would act more as a house of regions or provinces than it does in Australia (which is hardly at all in an explicit sense). Still, an elected Senate in Canada would significantly increase the representative and legislative capacity for achieving national consensus, by strengthening the political legitimacy of the federal government. Part of Australia's ability to achieve national strategies on energy and resource issues comes back to federal parliamentary consensus-building and

---

[16]The newly elected Liberal-National government has a clear majority in the lower House of Representatives but faces a minority position in the Senate.

legitimacy; and part is attributable to a political culture, on both sides of the major ideological divide, that does not shrink from the very words "national strategy." The obstacles to Senate reform in Canada remain significant, but the will to do something with the unreformed institution does seem to be quickening.

More formalized intergovernmental relations, on the other hand, would not necessarily strengthen the hand of the federal government, but would provide the means if not the will to achieve policy harmonization and national strategic outcomes in the many areas of federal-provincial interdependence. In the Australian model the states and territories are protected by the more formalized regime, and collective policy capacity is greatly enhanced through joint agencies and other mechanisms. Decision-making is more rapid and substantive. The political will to use the machinery waxes and wanes, and the more centralized legal and fiscal structure of the federal system forces the states and territories to be more cooperative with the federal government. Still, the appetite for effective collaboration seems to grow with the eating, and has survived a number of political changes in the federal and state capitals. For their part, Canadian governments are unlikely to depart from their deeply entrenched competitive culture, although that culture is clearly reinforced by the current federal government's approach of classical disentanglement ("open federalism"). While the need for wholesale change to more formalized intergovernmental cooperation would be hard to sell today (although one should recall the optimism that initially met the announcement ten years ago of the Council of the Federation; see IIGR 2003), a case can be made for selective improvements in some fields, particularly to engage the federal government in those areas where the provinces alone cannot do the job. Realistically, the prospects for such reform would be much better with the election of a federal government ready to interact more systematically with the provinces.

Turning to the specific cases of resource and energy issues, many Canadians have been concerned about the fragmented and undeveloped electricity market in Canada, and there is a need for an interregional grid in many provinces, if not one spanning the entire country. The obstacles to creating the Australian national market, given the states' control and ownership in the electricity sector, have been just as daunting as they are now in Canada, so many of Australia's institutional mechanisms for achieving a national market in whole or in part are clearly worth careful consideration. This is especially so in that the federal role in Australia has not been heavy-handed but rather facilitating. Similarly, although not discussed above, Australia also faces interstate pipeline issues, which intergovernmental relations also seem to have been overcoming successfully (although in this field the Canadian federal role and jurisdiction is more established).

On climate change policy, and greenhouse gas emissions regulation especially, the role of more effective intergovernmental institutions and processes as well as an elected Senate seem clear. These are not the only reasons why Australia is further ahead on these issues than Canada (or for that matter the United States), but are certainly contributing factors. The main lesson for Canada must be that we lack the certain capacity to resolve difficult regional and intergovernmental issues in a collaborative way. Too much relies on all the stars aligning; political consensus has to be overwhelming before even incremental progress can be made.[17] In short, the history of substantive intergovernmental cooperation on greenhouse gas emissions has been abysmal. Our governments should take another hard look at Australian (or European Union) models that incorporate co-decision, including through qualified majority voting, and the legislative establishment of joint, binding regulatory authority for the relevant ministerial council. A renewed Canadian commitment to strengthen intergovernmental institutions dealing with the GGE issues seems overdue.

Finally, on the fiscal issues, the solutions and process in Australia as a whole are not especially transferable to Canada, given our apparent preference for fiscally strong provinces. Still, we also face widening horizontal fiscal inequity as a result of the accumulation of resource rents in some provinces. The net operation of the GST distribution in Australia, whereby richer states get a smaller per capita entitlement, seems more transparent even if it is not wholly satisfactory to all parties. However, the calculation of fiscal capacity that goes into the distribution is famously complex and intrusive, even if it is done by a reputable independent agency (the Commonwealth Grants Agency). On the federal minerals tax, we have our own scarring history of the National Energy Program, which probably prevents any such proposal from achieving lift-off in Canada. A national carbon price/tax regime might be a very different case, in that it is not on its face confiscatory of resource wealth surpluses, but it would probably have regional distributive effects more profound than those in Australia and would have to be carefully balanced. All of which begs the question about whether a decentralized federal country can have "national" energy and resource strategies at all.

The clear lesson from Australia is that a country with so many similar conditions, and at least some similar institutions, successfully achieves national strategies as a matter of course, even if the direction and pace of national action can change abruptly depending on the federal party in power. We need not adopt their entire approach to find useful solutions. For our part, Canadians would be better served by at least starting with a franker discussion about what we might all gain from a national strategy or strategies on our most pressing energy and resource issues.

REFERENCES

ABS (Australian Bureau of Statistics). 2012. Accessed November 5, 2012. http://www.abs.gov.au/.

---

[17] For a broader comparative discussion on this point see my recent article (Brown 2012) and the other articles in the same special issue of the *Review of Policy Research*.

Anderson, George, ed. 2012. *Oil and Gas in Federal Systems*. Toronto: Oxford University Press.

Aulich, Chris, and Mark Evans, eds. 2010. *The Rudd Government: Australian Commonwealth Administration 2007–2010*. Canberra: ANU E Press.

Blackwell, Richard. 2013. "Down Under, True North on Diverging Paths." *Globe and Mail*, August 7, 2013.

Brown, Douglas M. 2002. *Market Rules: Economic Union Reform and Intergovernmental Policy-Making in Australia and Canada*. Montreal: McGill-Queen's University Press.

—. 2012. "Comparative Climate Change Policy and Federalism: An Overview." *Review of Policy Research* 29 (3): 322-33.

Brown, Douglas, with Herman Bakvis and Gerard Baier. 2011. "The Senate in Australia and Canada: Mr. Harper's Senate Envy and the Intrastate versus Interstate Debate." In *The Federal Idea: Essays in Honour of Ronald L. Watt*, edited by Thomas J. Courchene, John R. Allan, Christian Leuprecht, and Nadia Verrelli, 499-521. Montreal: McGill-Queen's University Press; Kingston: Institute of Intergovernmental Relations, Queen's University.

Cairns, Alan C. 1977. "The Governments and Societies of Canadian Federalism." *Canadian Journal of Political Science* 10 (4): 695-725.

CGC (Commonwealth Grants Commission). 2013. "Report on State Revenue Sharing Relativities – 2013 Update." Media release, Canberra.

Commonwealth of Australia. 2012. *GST Distribution Review, Final Report*. Canberra.

—. 2013. *Budget Paper No. 3: Australia's Federal Relations, 2013–14*. Canberra: Department of Treasury.

Coper, Michael. 1988. *Encounters with the Australian Constitution*. Sydney: CCH.

Coulombe, Serge. 2013. "Terms of Trade Changes, the Dutch Disease and Canadian Provincial Disparity." IIGR Working Paper 2013–01. Institute of Intergovernmental Relations, Kingston.

Courchene, Thomas J. 1996. "The Comparative Nature of Australian and Canadian Economic Space." In *Reforming Fiscal Federalism for Global Competition: A Canada-Australian Comparison*, edited by Paul Boothe, 7-21. Edmonton: University of Alberta Press.

Crowley, Kate. 2010. "Climate Clever? Kyoto and Australia's Decade of Recalcitrance." In *Global Commons, Domestic Decisions: The Comparative Politics of Climate Change*, edited by Katherine Harrison and Lisa Sundstrom, 201-28. Cambridge: MIT Press.

Economist. 2012a. "The Oil Barons Have a Ball: Natural Resources Are Not Really a Curse After All." February 18, 2012.

—. 2012b. "A Mining Tax for Australia: Your Tax or Mine?" March 24, 2012.

—. 2012c. "Australian Politics: Another Fine Mess." May 12, 2012.

—. 2012d. "Australia's Fiscal Reform: Tithes of Discontent." July 7, 2012.

—. 2012e. "Australia's Two-Track Economy: Hitched to China's Wagon." August 25, 2012.

—. 2012f. "The World in 2013: Booming On: Australia Pins Its Hope on China." December, 2012.

Fenna, Alan. 2007. "The Malaise of Federalism: Comparative Reflections on Commonwealth-State Relations." *Australian Journal of Public Administration* 66 (3): 298-307.

Galligan, Brian. 1995. *A Federal Republic: Australia's Constitutional System of Government*. Melbourne: Cambridge University Press.

Garnaut, Ross. 2008. *The Garnaut Climate Change Review: Final Report*. Melbourne: Cambridge University Press.

Garnett, Anne, and Phil Lewis. 2010. "The Economy." In *The Rudd Government: Australian Commonwealth Administration 2007–2010*, edited by Chris Aulich and Mark Evans, 181-98. Canberra: ANU E Press.

Gordon, David, and Douglas MacDonald. 2011. "Institutions and Federal Climate Change Governance: A Comparison of the Intergovernmental Coordination in Australia and Canada." Paper presented to the Canadian Political Science Association annual meetings, Waterloo, May 18, 2011.

Grant, Tavia. 2013. "Similarities, but 'Big Differences.'" *Globe and Mail,* May 8.

Harrison, Kathryn. 2011. "The Politics of Carbon Pricing in Australia, Canada and the United States." Paper presented to the American Political Science Association annual meetings, Seattle, September 3, 2011.

IIGR (Institute of Intergovernmental Relations, Queen's University). 2003. *Constructive and Cooperative Federalism: A Series of Commentaries on the Council of the Federation*. Kingston: IIGR.

Ker, Peter. 2013. "Mining Tax Revenue Slumps." *Sydney Morning Herald,* May 14.

Livingston, W.S. 1956. *Federalism and Constitutional Change*. Oxford: Clarendon.

Macintosh, Andrew, Deb Wilkinson, and Richard Denniss 2010. "Climate Change." In *The Rudd Government: Australian Commonwealth Administration 2007–2010*, edited by Chris Aulich and Mark Evans, 199-220. Canberra: ANU E Press.

Morris, Alan. 2007. "Commonwealth of Australia." In *The Practice of Fiscal Federalism*, edited by Anwar Shah, 43-72. Montreal: McGill-Queen's University Press.

Painter, Martin. 1998. *Collaborative Federalism: Economic Reform in Australia in the 1990s*. Melbourne: Macmillan.

Quiggan, John. 1996. *Great Expectations: Microeconomic Reform and Australia*. Sydney: Allen and Unwin.

Saunders, Cheryl. 2011. *The Constitution of Australia: A Contextual Analysis*. Oxford: Hart.

Smiley, Donald V., and Ronald L. Watts. 1985. *Intrastate Federalism in Canada*. Research Studies of the Royal Commission on the Economic Union and Development Prospects for Canada, vol. 39. Toronto: University of Toronto Press.

Victoria. 2011. *Budget 2011–12*. Melbourne: Department of Treasury.

Ward, Ian, and Randall Stewart. 2010. *Politics One*. Sydney: Palgrave Macmillan.

Watts, Ronald L. 1989. *Executive Federalism: A Comparative Analysis*. Kingston: Institute of Intergovernmental Relations, Queen's University.

—. 2008. *Comparing Federal Systems*. 3rd ed. Kingston: Institute of Intergovernmental Relations, Queen's University.

Western Australia. 2011. *Budget 2011–12*. Perth: Department of Treasury.

V

Identity and Resiliency

10

GUARDING THE NATION: RECONFIGURING CANADA IN AN ERA OF NEO-CONSERVATISM

Tim Nieguth and Tracey Raney

INTRODUCTION

In recent years, a number of media observers have suggested that the current federal government is attempting to rebrand Canadian national identity by emphasizing the military, war heroes, and the monarchy (Boesveld 2011; Martin 2010; Rowe 2011; Taber 2011). Such a shift would constitute a major change from previous constructions of the Canadian nation that rested on symbols of social equality, inclusiveness, and plurality, expressed in policies and documents such as universal health care, the Charter of Rights and Freedoms, and multiculturalism. Should the Conservative government's rebranding of the nation be successful, the consequences could be far-reaching. A reconfigured national identity tied to Canada's military and the monarchy might, for example, deepen ethno-national cleavages, erode cross-regional cohesion, and contribute to Conservative Party dominance. At the same time, the government's national policy agenda may encounter a number of roadblocks, including opposition from other political actors as well as public attachment to previous versions of national identity. It is therefore imperative to understand the scope of potential changes to Canada's national identity, as well as the political processes and mechanisms that enable or inhibit these changes.

In this chapter, we examine the nation-building strategies of the current Conservative government, and we argue that these strategies both extend and depart from previous constructions of Canadian nationhood in important ways. They are

---

Authors listed alphabetically.

an extension of previous strategies in that the Conservatives have continued the process of decoupling social policy from definitions of Canadian national identity, a process that began well before the party came to power in 2006. At the same time, the Conservative vision of the Canadian nation constitutes a significant departure from previous nation-building processes in at least two respects: it entails a change of the policy fields that are central to Canadian national identity, and it involves a redefinition of Canadian national symbols. This version of Canadian national identity is decidedly neo-conservative in its assumptions about the role of the state, market, and individual citizens in Canada's political community.

The chapter is divided into three sections. It will begin by discussing some of the key policy changes that are intended to transform Canada from a neo-liberal to a neo-conservative nation. In doing so, we focus on two elements of the Conservative nation-building strategy: the change of public policy fields centrally linked to Canadian national identity, and the reordering of symbols connected to Canadian nationhood. The second section of the chapter will examine some of the factors that have facilitated these changes, focusing on Canada's parliamentary system and the Conservative approach to Canadian federalism. Finally, the chapter considers some of the potential implications of recasting Canadian national identity along neo-conservative lines. These include the potential deepening of fault lines between Quebec, Aboriginal peoples, and the rest of Canada; increased regional divergence; an erosion of social citizenship; and changes to Canada's electoral landscape.

NATIONAL POLICY AND NEO-CONSERVATISM IN CANADA

The Canadian state has historically been central to the construction of the Canadian nation. In fact, Canada is sometimes described as a "state-nation" (Gwyn 1995; Stepan 2008) or "civic nation" (Ignatieff 1994; Nieguth 1997; Raney 2009). Although the Canadian state is clearly not the only actor involved in nation-building processes, it is equally clearly one of the most important participants in this enterprise. Modern states, including Canada's, command considerable resources, and they routinely employ them to help demarcate the limits of the communities they govern, create symbols that promote shared values and national mythologies, devise political institutions that structure political values and participation, and enact public policies that affect the status of different social groups and their place in the national community.

While many of these decisions may not explicitly be framed as an attempt to shape the nation, they have a cumulative impact on national identity. This impact can be understood through the lens of a national policy agenda, which Bradford (1998, 3) defines as an "overarching federal development strateg[y] for achieving economic growth and social cohesion within the Canadian political community."

National policy agendas provide broadly defined policy goals, or "overarching philosophical frameworks" that shape public policy (Bradford 1998, 3). Put differently, national policy agendas revolve around a relatively coherent set of political ideas that frame policy choices across a wide swath of public policy areas and determine which policy domains state actors do (or do not) perceive to be central to their vision of the nation.

The current Conservative government's national policy agenda is informed by neo-conservative ideology.[1] Broadly defined, neo-conservatism is a combination of neo-liberalism and social traditionalism (McBride and Shields 1993, 1997; Stelzer 2004; Teghtsoonian 1993). Neo-liberalism assumes that the free market is superior to the state in creating wealth and guaranteeing individual liberty; it aims to decrease state activity in the economy, and to increase state activity in the areas of safety, security, and the rule of law. Social traditionalism rejects the secular orientation of modern societies and the disintegration of the traditional roles of family, gender, religion, and morality. Social traditionalists believe that the state can and should be called upon to stop the erosion of traditional values. Overall, neo-conservatism advocates a leaner state, but not a weaker one – a state that limits social and economic redistribution, refrains from imposing a rights agenda that is at odds with traditional values, and focuses on the provision of security, law and order.

These ideological commitments appear in numerous texts and documents published by the Conservative Party and members of the Conservative government, including party platforms and policy declarations, speeches, and newspaper articles. They are conveniently summed up in a speech Prime Minister Harper delivered three years into his government's tenure; it is worth quoting this speech at some length:

> In the Canada of the future, we should be able to have one of the most free-enterprise, one of the most prosperous, societies on the planet. That would require us to govern according to conservative values. What exactly are those conservative values? [...] I like to summarize my idea of conservatism in three "Fs" – freedom, family and faith. Individual freedom, political and economic, is one of our fundamental values. It is absolutely critical. But it must be tempered. First, individual freedom must be tempered by family. We are part of a chain in which we honour and build upon those who came before us and in which we hope and look out for the future of those who will come after. Second, freedom must be tempered by faith that there is a right and wrong. It teaches us that freedom is not an end in itself, that how freedom is exercised

---

[1] This is not to suggest that all decisions made by the Conservative government will reflect neo-conservatism. There have been some Conservative government decisions that appear to depart from neo-conservatism, such as Mr. Harper's refusal to open up the same-sex marriage debate, and his decision to have a free vote in the House on abortion rights in the spring of 2012. These ideological departures are to be expected where public opinion is quite divided on particular issues, as is the case with same-sex marriage and abortion. Other factors that explain divergences from government policy are discussed later in the chapter.

matters as much as freedom itself. Freedom must be used well. And freedom can only be maintained if it is used well. (Harper 2009, A16)

As indicated above, the political thinking of leading Conservatives combines a belief in individual freedom and the virtues of the (well-regulated) marketplace with an insistence on the necessity of traditional social norms and forms of social organization. It thus weds some of the key elements of classical liberalism to central aspects of classical conservatism. This blend of ideas has exercised considerable influence over Conservative approaches to Canadian national identity, which we elaborate on below. It also builds on the neo-liberal legacy of previous governments, at the same time departing from that legacy in a number of crucial ways.

Retrenching the Welfare State

For at least the first three decades of the post-WWII era, the field of social policy played a central role in the articulation of Canadian nationalism. As Brodie (2002, 167) argues, the Keynesian welfare state gave new meaning to the Canadian nation, and to the idea that Canada was a "shared community of fate." This community was founded, in part, on the expectation that the Canadian state would play an important role in ensuring the social welfare of its citizens. Various state programs – including Unemployment Insurance (1941), Family Allowance (1945), universal Old Age Security (1951), and Medicare (1966) – established a social safety net that became the basis of Canada's collective identity. Social programs, including social transfers to the provinces, were also instrumental in maintaining national unity by evening out regional disparities across the country (Baker 1997; Banting 1987). During this time period, social policies provided much of the glue that bound the Canadian national community together (Béland and Lecours 2005, 198). Social policy was a foundational myth that helped carve out a welfare state nationalism that would define Canada from the 1940s to the 1970s and 1980s.

Over the last 30 years, the welfare state has become less important to Canada's national identity. Research highlights the extent to which the Canadian state became increasingly neo-liberal and securitized during this time (Brodie 2003; Lennox 2007; McBride and Shields 1993, 1997; Nimijean 2005). To be sure, this development began well before the current Conservative government took power in 2006 and was embraced by political parties other than the Conservatives. For example, in 1994, during the Liberal era, the Chrétien government invoked a program review and slashed overall government spending by 12 percent (or $14 billion) over a five-year period (Egan and Palmer 2011). During Prime Minister Chrétien's first year in office, federal government expenditures equaled 23.1 percent of Canada's GDP; by 2003, expenditures had dropped to 16 percent of GDP (Reynolds 2012).

The current Conservative government has continued the relative retrenchment of the state: in the 2011–12 fiscal year, federal spending amounted to 14.4 percent of the GDP; by 2015–16, it is projected to drop to 12.9 percent (Reynolds 2012). While absolute federal program expenses increased from $183 billion to $271 billion between 2002–03 and 2011–12 (Receiver General for Canada 2003, 2012), it is important to keep in mind that this increase occurred in the context of a rapidly growing and aging population, a global economic recession that resulted in considerably higher social expenditures, and a significant financial commitment to the war in Afghanistan. In addition, a substantial portion of this increase is owed to inflation, which amounted to 18 percent between 2003 and 2012. In the 2013–14 budget, total spending is expected to rise less than 1 percent from the 2012–13 budget. When inflation and population growth are considered, this constitutes an actual cut (Cheadle 2013). Direct program expenses (not including major transfers to the provinces and territories) are also expected to drop substantially (Cheadle 2013).

Social policy has been especially vulnerable to cutbacks over the last 20 years. For example, the first Chrétien budget in 1994 restrained federal expenditures and restructured social programming (Prince 2006, 214). This trend has continued under the Conservative government: in 2012, the Conservative government announced changes to Old Age Security (OAS), which will increase the age of eligibility from 65 to 67 between 2023 and 2029. As of January 2013, seasonal workers have to meet stricter criteria in order to remain eligible for Employment Insurance; in particular, the new criteria force them to travel farther for work at lower pay. Overall, according to the Parliamentary Budget Officer, the federal government's program spending (which consists of major transfers to persons or other levels of government, and direct program spending that includes money for social programs) is expected to remain lower than the historical average of 15.4 percent of GDP (Office of the Parliamentary Budget Officer 2012, 16). Long-term declines in social spending relative to GDP are also expected in social expenditure areas such as elderly benefits (the OAS, the Guaranteed Income Supplement, and the Allowance) and children's benefits (the Canada Child Tax Benefit and Universal Child Care Benefits). Projections for the Canada Health Transfer and Canada Social Transfer are also expected to decrease from 0.6 percent of GDP in 2011 to 0.3 percent in 2086 (Office of the Parliamentary Budget Officer 2012).[2]

Policy Fields and National Identity

In important ways, then, the Conservative government has continued Canada's retreat from the welfare state nationalism of previous generations. At the same

---

[2] The calculation for elderly benefits includes changes the Conservative government made to the age of eligibility to take effect 2023–2029. While elderly benefit spending would have reached 3.2 percent of GDP in 2036, the new rules of eligibility have lowered spending to 2.8 percent of GDP for the next 20 years.

time, there are compelling reasons to conclude that the Conservatives are reframing the Canadian nation along new lines that extend beyond the neo-liberal projects pursued by previous governments. Most importantly, this rebranding is not confined to the erosion of welfare state nationalism; in addition, it seeks to reset Canadian national identity by amplifying other elements of Canadian nationalism, including the monarchy, law and order, and the military.

In keeping with the principles of neo-conservatism, which prioritize safety and security, the Conservative government's national policy strategies cast the Canadian nation as imperilled, at risk, and vulnerable to attacks originating from inside and outside its borders. The 2011 party platform – entitled "Here for Canada" – emphasizes the party's intent to strengthen the armed forces and Coast Guard, crack down on human smuggling, deport foreign criminals, extend the security infrastructure program, combat terrorism, and protect Arctic sovereignty (Conservative Party of Canada 2011). In effect, the primary function of the state in this new, imperilled nation is to safeguard and secure its citizenry.

This emphasis on the "security state" is further reflected in a raft of government bills introducing new regulations and procedures in policy areas such as criminal justice, immigration, and citizenship. Bill C-10 (The Safe Streets and Communities Act; royal assent March 2012), an omnibus bill, contains myriad policy changes to the Criminal Code, Immigration and Refugee Protection Act, Controlled Drugs and Substances Act, Youth Justices Act, Crime Records Act, Corrections and Conditional Release Act, and State Immunity Act (Prince 2012, 63). Likewise, while social spending has been cut or frozen, federal spending on law and order has grown. Between 2006 and 2011, the federal government added approximately $2.5 billion in new expenditures on law enforcement and public order (Prince 2012, 62). Canada's suite of new crime legislation comes at the same time that Statistics Canada has reported a 20-year decline in police-reported crime in the country (Brennan 2012).

Similar shifts have transpired in Canadian foreign and defence policy. Since the 1960s, Canadian national identity has been closely tied to Canada's role on the international stage – including its position as a global middle power (Chapnick 2000) and a peacekeeping nation (Anker 2005). The current Conservative approach to the international stage is focused more squarely on the containment of threats to Canada and Canadian democracy through military means. First and foremost, the Conservative government's military emphasis can be seen in its increased defence expenditures. As a percentage of federal government expenditures, Canada's military spending has risen every year since the Conservative government took office. Overall, military expenditures increased from 6.3 percent in 2003 to 7.6 percent in 2010 (World Bank 2013).[3] In 2010–11, approximately $21.3 billion of

---

[3] Military expenditure includes all current and capital expenses related to the armed forces, including peacekeeping forces, defence ministries and other defence agencies, paramilitary,

total program spending was allocated to defence; this compares to $19.9 billion for Employment Insurance benefits, $11.2 billion for the Canada Social Transfer, and $26 billion for the Canada Health Transfer (Department of Finance Canada 2011).

The yearly increases to military spending came to an abrupt halt with the March 2012 budget; that budget included a 5 percent cut to the defence budget, which was to be phased in over a period of three years. Combined with spending cuts already announced in 2010, this will amount to approximately a 10 percent cut to defence spending. These deep cuts may shed some doubt on the centrality of the military to Conservative visions of Canadian national identity. However, it is important to take into account the impact of global economic and geopolitical contexts on this, previous, and future budgets. Two factors are particularly important in this respect: the global economic recession and a desire to maintain fiscal stability, and Canada's gradual withdrawal from the war in Afghanistan. Since entering that war in 2001, Canada incurred an estimated cost of $11.3 billion, or just over $1 billion per year (Government of Canada 2010). The 2012 budget was the first full budget delivered after the completion of Canada's combat mission in Afghanistan, and thus, in part, reflects the reduced cost of Canada's involvement in that particular conflict. It remains to be seen how the government's 2014 decision to deploy military aircraft, armed forces personnel, and additional military advisers to Iraq in order to combat ISIS will affect Canadian defence spending (Chase and Leblanc 2014; Sagan and O'Malley 2014).

In assessing the long-term significance of the military to the Conservative definition of Canadian identity, it is also important to keep in mind that military investments in the Arctic region (including Radarsat satellites, an armed ice-breaker, patrol vessels, and armed personnel) are proceeding steadily, if slowly. The Conservative government has explicitly tied these investments to Canadian national identity. In 2007, for example, Prime Minister Harper stated that "Canada has a choice when it comes to defending our sovereignty over the Arctic. We either use it or lose it. Make no mistake, this Government intends to use it. Because Canada's Arctic is central to our national identity as a northern nation. It is part of our history. And it represents the tremendous potential of our future" (cited in Lytvynenko 2011). The North has long occupied a central role in the Canadian national imagination (Grace 2001). What is new in the Conservative take on the North is its emphasis on securitization. When the Conservatives invoke the Arctic, it is not simply to reaffirm Canada's identity as a northern nation, but also to insist on the need to defend Canadian sovereignty and resource claims in the region. This line of reasoning helps provide a justification for a stronger military presence in Canada's Arctic region.

---

personnel, operations, procurement, research and development, and military aid. For further details see http://www.sipri.org/databases/milex/definitions.

Symbolic Reordering

As part of its rebranding strategy, the Conservative government has not only at-tempted to change the policy fields that are at the heart of Canadian national identity, but has simultaneously undertaken broad changes to Canada's symbolic order. Many of these policy changes highlight Canada's war history and heritage as a British colony. For example, the government invested significant effort and financial re-sources into commemorating the War of 1812. In contrast, the *Road to 2017* – the government's roadmap for celebrating the 150th anniversary of Confederation – downplays some of the key events and symbols that are commonly linked to rival visions of the Canadian nation, such as the 30th anniversary of the Charter of Rights and Freedoms, which took place with extremely limited government promotion in April 2012. In a similar vein, the government reinstated the "royal" designation for Canada's navy and air force in 2011, a move that emphasizes Canada's historical ties to the British monarchy. In 2012, the Bank of Canada began to release a new series of banknotes; among other changes, the redesigned bills place a stronger emphasis on war and the military.

In addition, the Conservative government published a new citizenship guide in 2009 (Government of Canada 2012). The citizenship guide is a particularly important means for transmitting national myths and symbols, since it is explicitly directed at newcomers to help them prepare for the citizenship test (applicants must pass this test in order to qualify for Canadian citizenship). In other words, the guide provides access to official membership in the nation; in doing so, it is an important opportunity for articulating national narratives that weave together the values, practices, institutions, and experiences considered central features of the Canadian national community. By extension, changes to the citizenship guide may point to larger changes in the symbolic order underpinning the national community. The fact that the Conservative government replaced the previous citizenship guide, released by the Liberal government under the title *A Look at Canada* (Government of Canada 1999), with a new guide entitled *Discover Canada*, accordingly merits some analysis.

A direct comparison shows that the vision of Canadian national identity en-trenched in these two documents contains important continuities, but that there are also several significant changes in the emphasis on different national symbols and experiences. According to Wilton (2010), *A Look at Canada* revolved around four major discourses on Canadian national identity: "(1) Canada is a nation of immigrants; (2) Canada is a country of regions; (3) Canada is a bilingual country; and (4) Canada is a multicultural country" (95). To a large extent, these themes also figure prominently in the new citizenship guide. However, *Discover Canada* differs from the earlier document in a number of respects. For example, where *A Look at Canada* makes scant mention of warfare, the military, or soldiers, *Discover Canada* is replete with military references: "war" and "warfare" are mentioned no less than 46 times (compared to two in *A Look at Canada*), "military" is used 11 times (compared to 1), and "soldiers" or "veterans" are mentioned 19 times (compared to 1).[4]

Military themes play an important role throughout the new citizenship guide, including the expanded discussion of citizenship responsibilities; in that section, aspiring Canadian citizens are informed that "[t]here is no compulsory military service in Canada. However, serving in the regular Canadian Forces [...] is a noble way to contribute to Canada and an excellent career choice" (Government of Canada 2012, 9). The guide's section on Canadian national symbols likewise makes repeated references to the military, the role of war in Canadian history, and the ways in which Canada honours its military personnel (Government of Canada 2012, 38-41).

In contrast to *A Look at Canada*, *Discover Canada* also puts much greater emphasis on Canada's ties to the British monarchy. Where the earlier document mentions the "Queen," "Crown," "Sovereign," or "Monarchy" only 13 times, the new guide uses these terms 53 times. Similarly, the Crown is the first item mentioned in *Discover Canada*'s section on national symbols, while it occupied a much less prominent place in the corresponding section of *A Look at Canada*. This stronger emphasis on the monarchy, like the increased attention paid to the military, resonates with the emphasis on social traditions and the "night watchman state" that are characteristic of neo-conservatism. The influence of neo-conservatism is also evident in the new citizenship guide's increased emphasis on law and order: while *A Look at Canada* mentions "law" 22 times, *Discover Canada* more than doubles this number (46).

At the same time, the new citizenship guide de-emphasizes symbols that are commonly associated with the vision of Canadian national identity championed by previous (Liberal) governments, such as the Charter of Rights and Freedoms and the environment. Thus, while *A Look at Canada* references the Charter – one of the key legacies of the Trudeau government – ten times, *Discover Canada* contains only three references to the Charter. The new citizenship guide also de-emphasizes environmental themes: while *A Look at Canada* mentioned the environment 13 times and dedicated an entire page to the environment, environmental protection, and sustainability, *Discover Canada* mentions the environment only eight times and does not pay sustained attention to this theme.

---

[4] These numbers (as well those provided in the following paragraphs) are based on a content analysis of the citizenship guides' main body; the analysis does not include mentions in the sample study questions, contact information, acknowledgments, or credits.

POLITICAL AND INSTITUTIONAL CONTEXTS

What factors have enabled the Conservative government to pursue such a comprehensive redefinition of Canadian national identity? The short answer is that there are several interrelated factors at work, rather than a single causal variable. These factors include, for example, a heightened emphasis on security in the post-9/11 era, a succession of global economic crises, Canada's relationship with the United States, the ascendancy of neo-liberalism since the 1970s, and the prime minister's control over the appointment of senators and Supreme Court judges. Given the scope of this chapter, we will devote our attention to two specific factors: Canada's style of Westminster parliament, and Conservative approaches to federalism and intergovernmental relations.

Westminster Parliament

It is commonplace to note that Canada's Westminster-style parliamentary system concentrates political power in the executive branch. In consequence, Parliament's ability to fulfill one of its core functions – control of the executive – is severely limited. While there is some debate about the scope of the Canadian prime minister's powers specifically (White 2005), there is relative consensus that these powers have increased over time (Savoie 1999, 2010; Simpson 2001). According to some political observers, the growth of prime ministerial influence has expanded since Stephen Harper assumed office in 2006 (Martin 2010). In 2013, *The Hill Times* voted Mr. Harper the most influential person on Parliament Hill for that year, describing him as the "central figure of the most centralized federal government in the country's history."

These developments make the prime minister an extraordinarily important figure in the Canadian political process, enabling him to use the policy-making machinery to enact policy changes broad and deep enough to reconstruct Canada's national identity. Recent years have seen a number of changes to the composition of the House of Commons that are likely to increase the prime minister's manoeuvrability on the national identity file. After six years of governing with a parliamentary minority, the Conservative party won a majority of seats in the 2011 federal election. This is significant for the development of Canadian national identity because it effectively ended the government's need to rely upon opposition parties to enact its legislative agenda.

The 2011 election may also have important consequences for national identity because it produced a new regional reality within the federal government: specifically, the Conservatives were able to win a majority of seats without the support of the province of Quebec. Of the 75 seats available in that province, the Conservatives managed to secure only five – the worst result for any (new or incumbent) governing party in at least 30 years. This continues a long-term decline of Quebec's representation within the governing party: despite small upswings in 1988 and 2000, the percentage of Quebec's seats won by the governing party has dropped from 99 percent in 1980 to a mere 7 percent in 2011. Figure 1 shows the percentage of regional seats won by governing parties since 1980. What is clear from the figure is that no governing party has managed to secure a parliamentary majority with such weak support from Quebec in recent decades; the closest comparison was the 1993 election, when the Bloc Québécois first emerged on the electoral scene. Even then, Prime Minister Chrétien won a majority with 25 percent of Quebec's seats.

Significantly, the Conservatives' declining fortunes in Quebec buck the trend emerging in other parts of the country. A comparison of the most recent election results to two other significant federal elections can serve to illustrate this point. In 1984, Brian Mulroney was able to stitch together a coalition of western populists, traditional Tories, and francophone nationalists. Figure 1 shows that in that year,

Figure 1: Percentage of Regional Seats Won by the Governing Party, 1980–2011

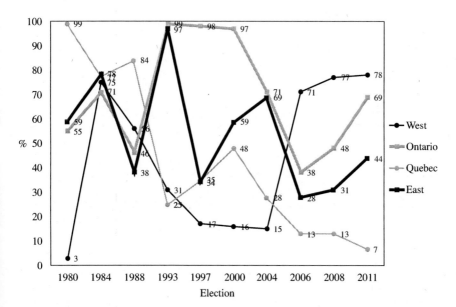

Notes: West: British Columbia, Alberta, Saskatchewan, Manitoba; East: Nova Scotia, New Brunswick, Prince Edward Island, Newfoundland and Labrador; territories not included.

Source: Parliament of Canada (2012).

regional representation in the Progressive Conservative governing caucus was quite high and somewhat balanced across the four regions (the East, Quebec, Ontario, the West): the governing party managed to win 78 percent (the East), 77 percent (Quebec), 71 percent (Ontario), and 75 percent (the West) of available seats, respectively. In the highly regionalized 1993 election, the Liberal governing party's share of seats dropped both in Quebec and in the West, but increased sharply in Ontario and the East. By comparison, in 2011, the governing Conservatives were able to increase their share of seats in Ontario, the West, and the East (continuing a trend established in the 2006 election), but the party lost seats in Quebec. Thus, for the first time in 30 years, electoral trends in Quebec are out of sync with those in the entire rest of the country.[5]

Changes to provincial representation in the House of Commons may ensure the continuation of this trend in subsequent federal elections. In December 2011, the Conservatives' Bill C-20 (the Fair Representation Act) received royal assent. This bill will alter the regional distribution of seats in the House for future elections; specifically, it increases the number of seats in Ontario by 15, Alberta by six, and British Columbia by six. Quebec will also receive three new seats, thus retaining 25 percent of seats. All in all, the size of the House will be increased by 30 seats. Bill C-20 has at least two major implications: first, it will make it easier for future governments to be formed in Ottawa without the support of the province of Quebec; and second, these changes will likely benefit the Conservative Party, since the West and Ontario are its two major regional support bases. In fact, had the new constituencies been in place for the 2011 federal election, the Conservatives would likely have picked up an additional 22 seats, increasing their share of seats from the 166 they won in 2011 to 188 (Chief Electoral Officer of Canada 2013, 10).

Open Federalism

Federalism (or, more precisely, the provinces) potentially limits the powers of the federal government. This has been especially true since WWII: in the postwar era, intergovernmental relations greatly intensified with the expansion of the welfare state and the emergence of several shared-cost social programs, such as health insurance (Banting 1987; Montpetit 2012). This intensification led to increased federal-provincial entanglement in a number of policy areas that have been central to Canada's postwar identity – welfare and health care, in particular. Historically, this development gave the provinces an important say in defining Canada's national identity by constraining the federal government's ability to unilaterally reconfigure the Canadian nation.

---

[5] Quebec's popular vote for the Conservative Party has also steadily declined, from 24.6 percent in 2006, to 21.7 percent in 2008, to 16.5 percent in 2011.

Yet, these constraints are being eroded by the current Conservative government in at least two ways, both of which weaken the role of the provinces on the national scene, reduce the necessity for federal-provincial negotiation, and increase the federal government's ability to implement its national policy agenda. As discussed earlier, the Conservative government has emphasized different policy fields in defining Canadian national identity than previous governments. While policy fields that fall under provincial jurisdiction (such as health care) have been de-emphasized, those that now take centre stage in defining the Canadian nation (such as national defence) fall within the exclusive purview of the federal government. One consequence of these changes is that the federal government can now more easily monopolize how the Canadian identity is defined, by whom, and through what means. In effect, a new federal bargain has been struck that grants more power to the premiers within their own jurisdictions, in exchange for reducing the provincial role in Canadian nation-building.

This new bargain is also reflected in the Conservative government's model of "open federalism," which seeks to disentangle the two orders of government through strict observance of the (narrowly construed) constitutional division of powers between Ottawa and the provinces. The principles of open federalism are laid out in some detail in Section D of the Conservative Party's 2005 "Policy Declaration." In brief, these principles include an adherence to the constitutional division of powers between Ottawa and the provinces; a belief in the importance of strong provinces; promises to limit the use of the federal spending power and to alleviate the "fiscal imbalance" between the two orders of government; support for strengthening instruments of interprovincial cooperation; and a commitment to address historical grievances of Quebec, the West, and Aboriginal peoples (Conservative Party of Canada 2005, 6-7).

It is worth emphasizing that open federalism does not simply aim to reduce the role of the federal government and strengthen that of the provinces. While it militates against federal encroachment in areas of provincial jurisdiction, the reverse also holds true. Thus, it would be misleading to describe open federalism as "de-centralizing"; as Young (2006, 10-11) points out, open federalism envisions strong provincial *and* federal governments. This is particularly important to discussions of national identity since open federalism effectively limits the capacity of provincial governments and their leaders to intervene in the development of federal policy in areas that are crucial to shaping national identity, as discussed above.

Open federalism arguably did not significantly affect the conduct of federal-provincial relations from 2006 to 2011 – notwithstanding Ottawa's recognition of Quebec as a "nation within a united Canada" in 2006, or increased provincial representation in certain areas of foreign policy. However, since winning a parliamentary majority in 2011, the Conservative government has made a number of high-profile decisions that suggest a rethinking of intergovernmental affairs along the lines of open federalism. For example, Jim Flaherty, the federal minister of finance, announced substantial changes to the Canada Health Transfer in December

2011. These changes were imposed unilaterally and without provincial consultation; unsurprisingly, they provoked vehement (but ultimately unsuccessful) protests from several provincial premiers.[6] More recently, Prime Minister Harper opted not to attend the 2012 First Ministers' Conference on economic issues – a decision that was widely interpreted as public affirmation that the federal government would limit its involvement in policy areas that fall under provincial jurisdiction.

POTENTIAL IMPLICATIONS

National Unity

A number of implications arise from the Conservative government's nation-building strategies that need to be considered. During the latter half of the twentieth century, Canada was mired in a protracted national unity crisis. For much of this time period, Quebec separatism posed a serious threat to the continuation of Confederation, despite the fact that the separatist option was soundly defeated in the first independence referendum of 1980. Less than a generation later, support for Quebec independence had reached a high water mark: in the 1995 referendum on independence, 49.5 percent of Quebec voters supported secession from Canada. The YES side (favouring secession) and NO side (opposing secession) were separated by a scant 54,000 votes.

Almost two decades after the second referendum, the picture has, in some ways, changed dramatically. While the separatist Parti Québécois managed to eke out an electoral victory in Quebec's 2012 provincial election, it received only 32 percent of the popular vote (3 percent less than in the previous election) and fell short of a legislative majority. Support for the Parti Québécois further declined in the 2014 provincial election, where it obtained 25 percent of the vote – a result that relegated the party to opposition status and put it a mere 2 percentage points ahead of the third place party, Coalition Avenir Québec. Similarly, surveys conducted in the lead-up to the 2012 election indicated that support for independence had dropped to a historic low of 28 percent (Ibbitson 2012; Mendleson 2012). Outside Quebec, political passions seem less exercised by the prospect of secession: a poll conducted in the summer of 2012 showed that 49 percent of respondents in the rest of Canada "don't really care" whether Quebec secedes (Ibbitson 2012). Canada's national unity crisis, while not resolved, no longer appears to dominate the political agenda of Canadian citizens in Quebec and the rest of Canada.

---

[6] These changes echo actions by some previous governments; in particular, the Chrétien government likewise made unilateral changes to the Canada Health Transfer, a step that provoked considerable resistance from the provinces.

On the basis of these observations, it might be tempting to conclude that sub-state nationalism is a much less potent and tumultuous political force in today's Canada. Such a conclusion is likely premature: while the national unity debates of the previous generation may have abated (at least for the time being), Quebec nationalism is by no means a spent force. The Conservative rebranding of Canada could potentially reinvigorate Quebec nationalism since it prioritizes a leaner state, emphasizes Canada's heritage as a British colony, and is less concerned with the goals of broader social inclusion. A newly assertive and decidedly more British national identity may further alienate not only Quebec, but also Aboriginal peoples and ethnic minorities who do not subscribe to a British-dominated view of Canada. A neo-conservative Canadian national identity may therefore further accentuate existing fault lines within Canadian society.

Federalism

The Conservative reconfiguration of national identity – and the institutional developments that allow this reconfiguration to take place – may also alter the workings of the Canadian federation, with far-reaching consequences for regional conflicts and national cohesion. Two issues bear special mention in this context. First, open federalism is characterized by increased separation of the federal and provincial orders of government. Instead of "federal-provincial diplomacy," interprovincial negotiations and intrastate federalism are left as the primary arenas of regional accommodation. The Conservatives' ongoing efforts at Senate reform broadly fit within this pattern, as does the intensification of interprovincial relations and cooperation in recent years (Berdahl 2011; Montpetit 2012).

If this pattern holds, it may have problematical consequences for national cohesion, since the interprovincial pillar of regional accommodation is likely to be brittle. For example, in the absence of strong federal leadership, provinces may have little incentive to adhere to national policy standards, encouraging policy divergence in areas that are central to the social and economic fortunes of individual Canadians, such as health care (Maioni 1999). Given the considerable fiscal imbalance between different provinces and varying degrees of dependence on federal transfers (McAllister 2011), policy divergence is a probable outcome. In addition, a reliance on interprovincial cooperation may strengthen regional cleavages (Berdahl 2011).

Second, the federal government has historically used its fiscal clout to offset some of the economic imbalances among the provinces. The Conservatives' commitment to state retrenchment (especially in areas of social policy and health care), in tandem with their philosophy of open federalism, may leave them less willing to do so. It is therefore perhaps not surprising that "the Harper government has, by both design and accident, reduced the redistributive impact of federal programs"

(McAllister 2011, 505). Recent measures, such as the changes to the Canada Health Transfer, may further accentuate regional differences, since these changes will lead to a gradual reduction of federal spending on health-care programs, relative to provincial spending. The long-term effects of open federalism will thus be a growing "fiscal gap" between more and less prosperous provinces.

Citizenship and Social Solidarity

The Conservative government's neo-conservative nation also redefines the roles of the state, market, national community, and individual citizens in Canada's political community. It assumes that the state has no interest in maintaining the social welfare of its citizens; rather, its role is to maximize market access and to oversee the protection, policing, and safeguarding of the nation. Importantly, in this national narrative, the state does not disappear, but reasserts itself in formidable ways, rising to guard the nation from both internal and external threats. In exchange for its protection, it demands individual responsibility and adherence to a closely scripted understanding of Canadian values as British, monarchical, and militarized.

This rebranding effort also redefines the bonds of citizenship and social solidarity; instead of collective rights and social equality, the emphasis is on individual responsibility to the state and to the wider community. As Prince (2012) argues, these are noteworthy shifts as the federal government moves away from defining Canadian nationhood as bounded by notions of "social cohesion, social investment, or social capital in social policy; or, in foreign policy, of middle power, human security and soft power discussed and promoted by previous Liberal administrations" (56). The Conservative government's rearticulation of the state from an instrument of "social investment" to a "nightwatchman" has shifted the nature of social citizenship from one bound together by the shared values of a "sharing, caring" nation to one bound by policing and security (Prince 2012).

This is not to say that the Conservative effort to rebrand Canadian national identity will inevitably be successful. Public opinion data on Canadian national identity indicate that a significant segment of the Canadian population continues to hold on to a more progressive, liberal definition of Canada tied to health care, the Charter of Rights and Freedoms, multiculturalism, and peacekeeping (Graves and Valpy, this volume). If this is the case, it is entirely possible that competing visions of identity within the rest of Canada will challenge the neo-conservative national narrative. This dynamic may produce further social discord and weaken Canada's "shared community of fate" in future generations. How and whether these perspectives will be accommodated within the nation-building strategies of the Conservative government will be crucial to Canadian social cohesion and solidarity.

Electoral Consequences

The Conservative government's changes to Canada's national identity may also have electoral consequences. The 2011 election shows that it is possible to win a majority government in Ottawa without substantial support from Quebec. This is a lesson the Conservative Party is likely to take to heart as it continues to build its electoral coalition outside Quebec. As Flanagan (2011, 104) notes, this is a concerted strategy: as far back as 1996, Mr. Harper believed that the formation of a conservative coalition similar to that of Mr. Mulroney's was both possible and necessary in order to form government. When the Conservatives failed to make satisfactory inroads with the Quebec electorate in 2006 and 2008, they jettisoned that element of the Mulroney coalition and tried instead to shore up their support in ethnic minority communities. The electoral need to win seats in Quebec has, in the past, limited the extent to which the government party could articulate a strong Canadian national identity that was at odds with the values of a significant portion of the Quebec electorate. Mr. Harper's new conservative coalition – based firmly on the West and Ontario – frees him from this constraint. Without Quebec, the Conservatives have had greater leeway to pursue a more explicit Canadian identity policy agenda.

In addition, the rebranding of Canadian national identity may – if successful – further cement the dominance of the Conservative Party and undermine the prospects of a Liberal return to power. In 2011, Prime Minister Harper delivered a speech in Calgary that celebrated the end of Liberal hegemony and the ascendance of the Conservative Party, emphasizing the role of political values and ideas in bringing about this development: "Conservative values are Canadian values. Canadian values are conservative values. They always were ... and Canadians are going back to the party that most closely reflects who they really are: the Conservative Party, which is Canada's party" (cited in McCarthy 2011). Research suggests that national identities help guide and shape citizens' political perceptions, behaviours, and policy positions (Raney and Berdahl 2011). Should the new, neo-conservative national identity resonate with significant segments of the Canadian electorate, this may prove Mr. Harper's assessment correct and entrench the Conservative government's grip on power. In consequence, the effort to rebrand Canada may generate significant electoral rewards for the Conservatives. However, that effort also poses considerable strategic risks. As already mentioned, a large number of Canadians do not appear to support the Conservative vision of Canada, and so the opposition parties could conceivably use the Conservatives' attempt to rebrand Canadian identity as a wedge issue in future federal elections.

CONCLUSION

As discussed above, the Conservative government's national policy agenda entails a change of policy fields deemed relevant to the neo-conservative nation. It eschews the welfare state nationalism of previous generations and seeks to build a national identity that emphasizes the military, law and order, and traditional national symbols like the monarchy. The symbolic reordering of Canadian national identity has followed broadly similar lines, underscoring themes closely connected to the Conservative Party's ideological commitments, and downplaying symbols associated with a Liberal vision of Canada. These changes were facilitated at least in part by the easing of various constraints on the federal government, providing space for the emergence of a new brand of Canadian nationalism – one that builds on 30 years of neo-liberal ascendancy.

The success of these rebranding efforts, however, is not a foregone conclusion; there are a number of factors that may complicate the full restructuring of Canadian national identity along neo-conservative lines. First, political parties are strategic actors; as such, they frequently adapt their policy agendas to political necessity and electoral expediency. As a result, the government's concrete policy decisions may not always readily reflect its commitment to neo-conservatism. For instance, at the 2010 G8 summit, Prime Minister Harper pledged to provide $2.85 billion for a global maternal and child-health initiative between 2010 and 2015 (Payton 2011). While this financial commitment has been lauded by many global leaders, the implementation of this initiative was limited by socially conservative views on women's reproductive rights. Initially, the Conservatives announced that they would prohibit funding to groups that provide family planning and safe abortion services; this decision was later revised to ban funding for only pro-choice groups (Cheadle 2011).

Second, governments, like other social actors, operate in a context that is not of their choosing. Changing circumstances at the international and domestic level can enhance or limit the extent to which the government is able to implement its national policy agenda. For example, the transition to a neo-conservative national identity was partly facilitated by the events of 9/11 and the increased emphasis on security in many Western nations. Similarly, global economic crises – such as the one currently being played out in the Eurozone – affect the Conservative government's policy choices.

Finally, national identity construction is an exceedingly complex process that involves a large and rather dissonant cast of actors (including the state, opposition parties, interest groups, religious leaders, the mass media, and individual citizens). In Canada, many of these actors subscribe to visions of national identity that conflict with the neo-conservative brand. The presence of countervailing perspectives may temper the speed with which wholesale reinterpretations of Canadian identity are introduced or are ultimately successful. Given the complexity of national identity processes generally, what remains surprising is how successfully the Conservative government has reconfigured Canadian identity thus far over a relatively short period of time.

REFERENCES

Anker, Lane. 2005. "Peacekeeping and Public Opinion." *Canadian Military Journal* 6 (2): 23-32.

Baker, Maureen. 1997 (June). "The Restructuring of the Canadian Welfare State: Ideology and Policy." Social Policy Research Centre. SPRC Discussion Paper No. 77. Accessed February 23, 2013. http://nchsr.arts.unsw.edu.au/media/File/dp077.pdf.

Banting, Keith. 1987. *The Welfare State and Canadian Federalism*. Kingston and Montreal: McGill-Queen's University Press.

Béland, Daniel, and André Lecours. 2005. "Nationalism and Social Policy in Canada and Quebec." In *The Territorial Politics of Welfare*, edited by Nicola McEwan and Luis Moreno. New York: Routledge.

Berdahl, Loleen. 2011. "Region-Building: Western Canadian Joint Cabinet Meetings in the 2000s." *Canadian Public Administration* 54 (2): 255-75.

Boesveld, Sarah. 2011. "Canada's Royal Rebrand." *National Post*, September 10. Accessed November 27, 2012. http://news.nationalpost.com/2011/09/10/canada%E2%80%99s-royal-rebrand.

Bradford, Neil. 1998. *Commissioning Ideas: Canadian National Policy Innovation in Comparative Perspective*. Toronto: Oxford University Press.

Brennan, Shannon. 2012. "Police-Reported Crime Statistics in Canada, 2011." Statistics Canada, July 24. Accessed February 20, 2013. http://www.statcan.gc.ca/pub/85-002-x/2012001/article/11692-eng.pdf.

Brodie, Janine. 2002. "Elusive Search for Community: Globalization and the Canadian National Identity." *Review of Constitutional Studies* 7 (2): 155-78.

—. 2003. "On Being Canadian." In *Reinventing Canada: Politics of the 21st Century,* edited by Janine Brodie and Linda Trimble. Toronto: Prentice Hall.

Chapnick, Adam. 2000. "The Canadian Middle Power Myth." *International Journal* 55 (20): 188-206.

Chase, Steven, and Daniel Leblanc. 2014. "Additional Canadian Troops Deployed in Iraq." *Globe and Mail*, October 6. Accessed November 4, 2014. http://www.theglobeandmail.com/news/politics/ottawas-debate-on-mission-to-fight-islamic-state-gets-underway/article20943294.

Cheadle, Bruce. 2011. "Harper Heralds Maternal Health's Rosy Future as UN Meeting Opens." *Globe and Mail*, January 26. Accessed February 20, 2013. http://www.theglobeandmail.com/news/politics/harper-heralds-maternal-healths-rosy-future-as-un-meeting-opens/article563641.

—. 2013. "Flaherty Budget Puts Brakes on Spending Hoping to Accelerate Growth." *Macleans.ca*, March 21. Accessed April 30, 2013. http://www2.macleans.ca/2013/03/21/flaherty-budget-puts-brakes-on-spending-hoping-to-accelerate-growth-2.

Chief Electoral Officer of Canada. 2013. *Transposition of Votes: 2013 Representation Order*. Ottawa: Elections Canada.

Conservative Party of Canada. 2005. "Policy Declaration." Accessed November 16, 2012. http://www.oshawaconservative.ca/PDF/050319%20CPC%20Policy%20Declaration.pdf.

—. 2011. "Here for Canada." 2011 Party Platform. Accessed February 18, 2013. http://www.conservative.ca/media/2012/06/ConservativePlatform2011_ENs.pdf.

Department of Finance Canada. 2011. *Your Tax Dollar*. Accessed February 26, 2013. http://www.fin.gc.ca/tax-impot/2011/2011-eng.pdf.

Egan, Louise, and Randall Palmer. 2011. "The Lesson from Canada on Cutting Deficits." *Globe and Mail Online*, November 21. Accessed February 24, 2013. http://www.theglobeandmail.com/report-on-business/economy/the-lesson-from-canada-on-cutting-deficits/article4252006/?page=all.

Flanagan, Tom. 2011. "The Emerging Conservative Coalition." *Policy Options* (June/July): 104-8.

Government of Canada. 1999. *A Look at Canada*. Ottawa: Minister of Public Works and Government Services Canada.

—. 2010. "Canada's Engagement in Afghanistan." Accessed November 15, 2012. http://www.afghanistan.gc.ca/canada-afghanistan/news-nouvelles/2010/2010_07_09.aspx?view=d.

—. 2012. *Discover Canada: The Rights and Responsibilities of Citizenship*. Ottawa: Minister of Citizenship and Immigration Canada.

Grace, Sherrill E. 2001. *Canada and the Idea of North*. Montreal and Kingston: McGill-Queen's University Press.

Gwyn, Richard. 1995. *Nationalism without Walls: The Unbearable Lightness of Being Canadian*. Toronto: McClelland and Stewart.

Harper, Stephen. 2009. "Off Liberty." *National Post*, April 30, A16.

Hill Times. 2013. "He's Got the Power: Prime Minister Stephen Harper Most Influential in 2013." *Hill Times*, January 8. Accessed January 28, 2013. http://www.hilltimes.com/100-most-influential/2013/01/28/he%E2%80%99s-got-the-power-prime-minister-stephen-harper-most-influential-in/33513.

Ibbitson, John. 2012. "Do We Care If Another Tussle for Quebec Sovereignty Happens Now?" *Globe and Mail*, August 31. Accessed November 20, 2012. http://www.theglobeandmail.com/news/politics/elections/do-we-care-if-another-tussle-for-quebec-sovereignty-happens-now/article4513853.

Ignatieff, Michael. 1994. *Blood and Belonging: Journeys into the New Nationalism*. New York: Farrar, Straus, and Giroux.

Lennox, Patrick. 2007. "From Golden Straightjacket to Kevlar Vest: Canada's Transformation to a Security State." *Canadian Journal of Political Science* 40 (4): 1017-38.

Lytvynenko, Anneta. 2011. "Arctic Sovereignty Policy Review." Paper prepared for the Ad Hoc Committee of Deputy Ministers on the Arctic, April 5. Accessed November 3, 2012. http://www4.carleton.ca/cifp/app/serve.php/1355.pdf.

Maioni, Antonia. 1999. "Decentralization in Health Policy: Comments on the ACCESS Proposal." In *Stretching the Federation: The Art of the State in Canada*, edited by Robert Young. Kingston: Institute of Intergovernmental Relations.

Martin, Lawrence. 2010. *Harperland: The Politics of Control*. Toronto: Viking Canada.

McAllister, James A. 2011. "Redistributive Federalism: Redistributing Wealth and Income in the Canadian Federation." *Canadian Public Administration* 54 (4): 487-507.

McBride, Stephen, and John Shields. 1993. *Dismantling a Nation: Canada and the New World Order*. Halifax: Fernwood.

—. 1997. *Dismantling a Nation: The Transition to Corporate Rule in Canada*. 2nd ed. Halifax: Fernwood.

McCarthy, Shawn. 2011. "Harper's Arrogance Will Come Back to Haunt Him, Opposition Says." *Globe and Mail*, July 10. Accessed January 19, 2013. http://www.theglobeandmail. com/news/politics/harpers-arrogance-will-come-back-to-haunt-him-opposition-says/ article2092763.

Mendleson, Rachel. 2012. "Quebec Election 2012: Economic Turmoil around World Makes Sovereignty Threat a (Relatively) Minor Problem." *Huffington Post*, September 4. Accessed November 20, 2012. http://www.huffingtonpost.ca/2012/09/04/quebec-economy -election_n_1853676.html.

Montpetit, Eric. 2012. "Are Interprovincial Relations Becoming More Important Than Federal-Provincial Ones?" *Federal News* 3 (6): 1-6.

Nieguth, Tim. 1997. *English Canadian Concepts of the Nation: Perspectives on Contemporary Developments*. Augsburg: Institute for Canadian Studies.

Nimijean, Richard. 2005. "Articulating the 'Canadian Way': And the Political Manipulation of the Canadian Identity." *British Journal of Canadian Studies* 18 (2): 26-52.

Office of the Parliamentary Budget Officer. 2012. *Fiscal Sustainability Report 2012*. September 27. Accessed November 15, 2012. http://www.pbo-dpb.gc.ca/files/files/ FSR_2012.pdf.

Parliament of Canada. 2012. "House of Commons Regional Representation: 1867 to Date." Last modified November 12, 2012. http://www.parl.gc.ca/Parlinfo/Compilations/ HouseOfCommons/RegionalRepresentation.aspx?Menu=HOC-Representation.

Payton, Laura. 2011. "Maternal and Child Health Projects Get $82M from Canada." *CBC News.ca*, September 20. Accessed February 20, 2013. http://www.cbc.ca/news/health/ story/2011/09/20/pol-harper-maternal-health.html.

Prince, Michael J. 2006. "La Petite Vision, Les Grands Decisions—Chrétien's Paradoxical Record in Social Policy." In *The Chrétien Legacy: Politics and Public Policy in Canada*, edited by Lois Harder and Steve Patten. Montreal and Kingston: McGill-Queen's University Press.

—. 2012. "The Hobbesian Prime Minister and the Night Watchman State: Social Policy under the Harper Conservatives." In *How Ottawa Spends—The Harper Majority, Budget Cuts, and the New Opposition*, edited by G. Bruce Doern and Christopher Stoney. Montreal and Kingston: McGill-Queen's University Press.

Raney, Tracey. 2009. "As Canadian as Possible...Under What Circumstances? Public Opinion on National Identity in Canada outside Québec." *Journal of Canadian Studies* 43 (3): 5-29.

Raney, Tracey, and Loleen Berdahl. 2011. "Shifting Sands? Citizens' National Identities and Pride in Social Security in Canada." *American Review of Canadian Studies* 41 (3): 259-73.

Receiver General for Canada. 2003. *Public Accounts of Canada 2003*. Ottawa: Minister of Public Works and Government Services.

—. 2012. *Public Accounts of Canada 2012*. Ottawa: Minister of Public Works and Government Services.

Reynolds, Neil. 2012. "'Incredible Shrinking Government' Message More Important Than Ever." *Globe and Mail*, September 6. Accessed February 13, 2013. http://m.theglobeandmail .com/report-on-business/rob-commentary/incredible-shrinking-government-message-more-important-than-ever/article535380/?service=mobile.

Rowe, Matthew. 2011. "Monarchy Lies at Heart of Our Identity." *Hill Times*, September 19. Accessed November 27, 2012. http://www.hilltimes.com/opinion-piece/politics/2011/09/19/monarchy-lies-at-heart-of-our-identity/28130.

Sagan, Aleksandra, and Kady O'Malley. 2014. "ISIS Mission: MPs Approve Canada's Air Combat Role." *CBC News*, October 7. Accessed November 4, 2014. http://www.cbc.ca/news/politics/isis-mission-mps-approve-canada-s-air-combat-role-1.2790189.

Savoie, Donald J. 1999. *Governing from the Centre: The Concentration of Power in Canadian Politics*. Toronto: University of Toronto Press.

—. 2010. *Power—Where Is It?* Montreal and Kingston: McGill-Queen's University Press.

Simpson, Jeffrey. 2001. *The Friendly Dictatorship*. Toronto: McClelland and Stewart.

Stelzer, Irwin, ed. 2004. *Neoconservatism*. London: Atlantic Books.

Stepan, Alfred. 2008. "Comparative Theory and Political Practice: Do We Need a "State-Nation" Model as Well as a 'Nation-State' Model?" *Government and Opposition* 43 (1): 1-25.

Taber, Jane. 2011. "Harper Spins a New Brand of Patriotism." *Globe and Mail*, August 19. Accessed February 24, 2013. http://www.theglobeandmail.com/news/politics/ottawa -notebook/harper-spins-a-new-brand-ofpatriotism/article2135876.

Teghtsoonian, Katherine. 1993. "Neo-conservative Ideology and Opposition to Federal Regulation of Child Care Services in the United States and Canada." *Canadian Journal of Political Science* 26 (1): 97-121.

White, Graham. 2005. *Cabinets and First Ministers*. The Canadian Democratic Audit Series. Vancouver: UBC Press.

Wilton, Shauna. 2010. "State Culture: The Advancement of 'Canadian Values' among Immigrants." *International Journal of Canadian Studies* 42: 91-104.

World Bank. 2013. "Military Expenditure (% of Central Government Expenditure)." Accessed February 14, 2013. http://data.worldbank.org/indicator/MS.MIL.XPND.ZS.

Young, Robert. 2006. "Open Federalism and Canadian Municipalities." In *Open Federalism: Interpretations, Significance*, edited by Keith Banting. Kingston: Institute of Intergovernmental Relations.

11

BEING CANADIAN TODAY:
IMAGES IN A FRACTURED MIRROR

Frank Graves, Jeff Smith, and Michael Valpy

To begin at the beginning: Who is us? Where is here? What is the personality of our imagined community on the top half of the North American continent, this question Canadians perpetually pose to themselves out of the DNA in their bones?

Seventy years ago, journalist-mythologizer Bruce Hutchison (1948), in *The Unknown Country: Canada and Her People*, lyrically asked: "Who can know our loneliness, on the immensity of prairie, in the dark forest and on the windy sea rock? A few lights, a faint glow is our largest city on the vast breast of the night, and all around blackness and emptiness and silence, where no man walks" (3). Lovely, a song from Canada's young adult years, but no hint now to our identity as we tread, warily and in disparate clumps, into the twenty-first century. EKOS Research for many years has been analyzing and recording how Canadians see themselves, their values and their identities. For the 2012 State of the Federation conference at Queen's University, it was asked to update a presentation it did on the topic in the late nineties (Graves 1999). What emerges is a picture both fascinating and disturbing and, without being unnecessarily alarmist, we believe that possibly never before in our history has it been so necessary to ask who is us and where is here.

There are alien economic, political, and demographic stresses on Canadian society for which the country has developed few remedies, certainly none to match the dimensions of what it faces. The stresses limn a portrait of profound changes in Canada connected to shifts in the economy, shifts in demographics, shifts in class structure and in perceptions of government and the role of the state.

It is a portrait of a society more polarized and riven than what we've known in decades if indeed there is precedent. It reveals that trust in democracy has reached a historic low (Figure 1). It shows Canadians to be increasingly pessimistic about their prospects in the new global economic order, a people fearing the end of progress in their and their children's lives. It shows their confidence in their country's direction to be declining sharply (Figure 2), especially among those under 40, the university educated, and Quebecers.

Yet while Canadians are raising questions about their journey together and, in fact, whether they still are journeying together, perhaps paradoxically their attachment to their country (while modestly declining) remains high across all demographic cohorts, higher than national attachment in virtually every other country on Earth (Figure 3). And surprisingly (if not equally paradoxically), the values of most Canadians are at odds with the social conservative convictions of the political right – certainly of their federal government – and the maxims of the political marketplace that declare as a given the need to advocate a minimalist state as the only certain path to electoral success.

Figure 1: Health of Democracy

Q. How would you rate the overall health of democracy at the federal level in Canada?

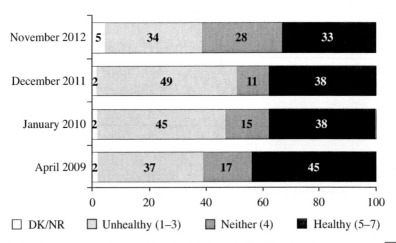

Source: EKOS Research Associates (2013c, 1).

Figure 2: Direction of Government

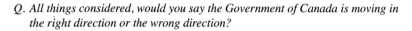

Q. All things considered, would you say the Government of Canada is moving in the right direction or the wrong direction?

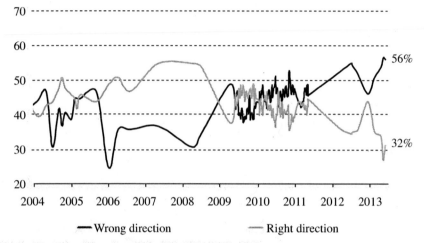

—— Wrong direction   —— Right direction

Note: This study was conducted using Interactive Voice Response (IVR) technology, which allows respondents to enter their preferences by punching the keypad on their phone, rather than telling them to an operator. The field dates for this survey are May 22–26, 2013. In total, a random sample of 3,318 Canadian adults aged 18 and over responded to the survey. The margin of error associated with the total sample is +/-1.7 percentage points, 19 times out of 20.

Source: EKOS internal survey.

These contradictory trends make Canadians a puzzling people, with an image in the mirror difficult to decode. Whatever our "us," it is beyond easy grasp of social scientists and journalists. "All about us lies Canada," wrote Hutchison the mythologizer, "forever untouched, unknown, beyond out grasp, breathing deep in the darkness.... They could not know us, the strangers, for we have not known ourselves" (1948, 3). And yet – who knows? – our attachment to the country and, fingers crossed, the resilience of our supposed innate Hegelian communitarianism may be the talismans we need to get us safely through danger to the other side.

Figure 3: Personal Sense of Belonging

Q. How strong is your own personal sense of belonging to ...

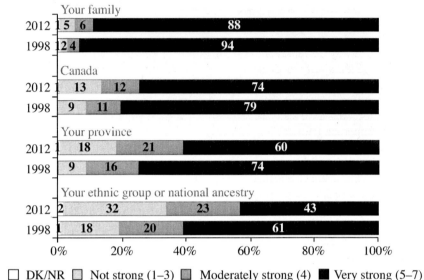

☐ DK/NR ☐ Not strong (1–3) ▨ Moderately strong (4) ■ Very strong (5–7)

BASE: Canadians; most recent data point November 20–29, 2012 (n=1,181).
Copyright 2013. No reproduction without permission.

Source: EKOS Research Associates (2013b, 16).

FRACTURES AND POLARIZATION (AND THE STATE AS IDENTITY)

Michael Ignatieff (2009), when he was leader of the Liberal Party of Canada, wrote:

> We need a public life in common, some set of reference points and allegiances to give us a way to relate to the strangers among whom we live. Without this feeling of belonging, even if only imagined, we would live in fear and dread of each other. When we can call the strangers citizens, we can feel at home with them and with ourselves. Isaiah Berlin described this sense of belonging well. He said that to feel at home is to feel that people understand not only what you say, but also what you mean. (4)

If Ignatieff's fellow citizens embraced those words, it was not to his political benefit. Indeed, in EKOS's findings, Canadians' public life in common is beset by boulders.

This chapter explores the shifts that are impacting Canadian society. It examines the reasons for those shifts and the deeper and sharper fractures in Canadian society that they are causing. It assesses what effect those fractures might have on Canadians' sense of being together – their social cohesion – and what remedies might be found for the good of the Canadian collective journey.

The central thesis is twofold: one, that Stephen Harper's ruling Conservatives are going one way on values and notions of government and the role of the state, while the great majority of Canadians are heading in the opposite direction; and, two, that almost all Canadians – primarily middle-class and young Canadians – are afraid, very afraid, for their economic future.

Most Canadians have dismissed social conservative values from their catalogue of what's important to them. Where there is movement on values, it is *away* from the right although, oddly, most *believe* the country is becoming a more socially conservative society. It's not. In addition, more and more Canadians are rejecting the idea of a minimalist state, likely not coincidentally with their fears for the economic future and the precarious workplace that is being fashioned by globalized capitalism with the support of the Canadian government and the governments of many other advanced industrialized societies. This drift away from the values embraced by the Conservatives is led by the younger generations (where the decline in attachment to social conservative values is most striking), as well as by Quebecers and metropolitan and university-educated Canadians. Many of the core values that define the Conservative base are almost irrelevant to these four groups. And while researchers at Heritage Canada who looked into the state of social cohesion more than a decade ago found growing age and education fractures on values (Jeannotte et al. 2002), the fault lines have sharpened over time as a result of Canadians' growing polarization. In less than a decade the percentage of Canadians calling themselves non-ideological has shrunk from 50 percent to 30 percent, leaving 70 percent polarized between small-c conservative and small-l liberal. The polarization corresponds with the emergence of a right-wing federal government. A values shift was not the causal agent – although the appearance of a political shift to the right is a product of the political success of a right-wing federal government supported by a minority. Whether the single-member district plurality electoral system is responsible for translating these fault lines into patterns of political representation is not known, although it may be somewhat the case in Alberta and Quebec. In any event, there are other, more plausible answers.

Polarization is taking place around the role and power of the state, foreign policy, civil rights versus national security, austerity versus social investment and the fears of economic insecurity. It is also taking place around democratic institutions: the overwhelming majority of younger Canadians are withdrawing from formal democratic participation, albeit a trend that began before Stephen Harper's Conservatives took power.

The identity of Canadians as connected to their image of the state may not be obvious at first glance. However, commentators such as Richard Gwyn (1995) have argued (and we concur) that Canadians' identity beginning with Confederation has been forged by what he called "state nationalism," a dialogue of citizens linked to their national institutions such as medicare, military peacekeeping and diplomacy, national railroads, the Charter of Rights and Freedoms, and the CBC.

Thus, it is interesting to consider how national identity might be evolving in an era where both the state's role and its relative size have taken a decidedly different trajectory in Canadians' lives. What we see, with the younger generation of adults, is that their identity – up until recently, at any rate – is not driven by what governments do but by how Canadians live, an ethos unfolding daily at interpersonal and societal levels. Among older Canadians, on the other hand, we see a fault line between those who continue to accept the state as the common terrain on which Canadians meet to defend each other's class interests and those who see the state as a barrier to their prosperity and even their liberty. And what EKOS finds, as shown in Figure 4, is a sharp decline among those listing "minimal government intrusions" as an important value in their lives, a substantial gap between younger and older Canadians on the issue (16 points, compared to a statistical tie in 1994), and a considerable gap between university-educated and college-educated Canadians (15 points, a gap that did not exist at all in 1994).

Figure 4: Importance of Minimal Government Intrusion

*Q. If you were to direct Canadian society as to which goals or values should be most
important in its direction, how important would you say each of the following goals
and values should be?* ***Minimal government intrusions***

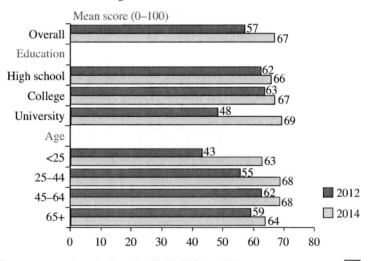

Source: EKOS Research Associates (2013b, 35).

ECONOMY, INSECURITY, INEQUALITY, AND THE WORKPLACE

Economic changes are powerfully reshaping how Canadians see their culture and country. As affluence and power have shifted westward, so have the peaks of national attachment and confidence in national direction, at least until Alberta's economy ran off the Bitumen Cliff. Nonetheless, overall the national outlook, as shown in Figure 5, is heavily affected by a growing fear that progress – the premise that collective and individual economic output would continue to rise forever, fattening social services, entitlements, and anticipated comfortable well-being and promising greater opportunity for each new generation – is no longer a guaranteed by-product of effort and skill. EKOS finds a large majority of Canadians to be both fearful of their relative

Figure 5: Direction of Country

Q. All things considered, would you say the country is moving in the right direction or the wrong direction?

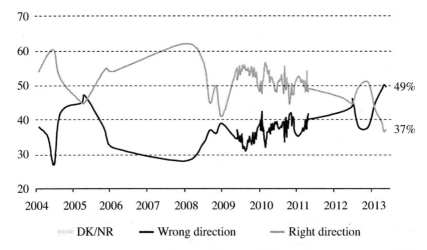

BASE: Canadians (half-sample); most recent data point May 22–26, 2013 (n=1,699).
Copyright 2013. No reproduction without permission.

Note: This study was conducted using Interactive Voice Response (IVR) technology, which allows respondents to enter their preferences by punching the keypad on their phone, rather than telling them to an operator. The field dates for this survey are May 22–26, 2013. In total, a random sample of 3,318 Canadian adults aged 18 and over responded to the survey. The margin of error associated with the total sample is +/-1.7 percentage points, 19 times out of 20.

Source: EKOS internal survey.

decline in a new global economic order and increasingly resentful of a new class ordering seen as allocating the lion's share of the meagre growth the economy is generating to an extremely narrow cadre of privileged Canadians (Figures 6 and 7).

Figure 6: Long-Term Global Economic Outlook

Q. As you may know, many Western nations are facing serious debt problems and global economic growth has stagnated over the last five years. Some economists say future generations will call this period "the lost decade." Do you believe that we are heading towards a period of prolonged economic stagnation or do you believe that the world economy will begin to improve in the near future?

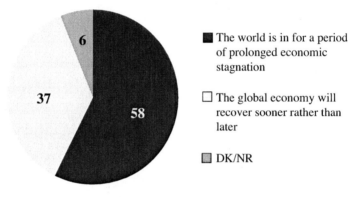

■ The world is in for a period of prolonged economic stagnation

☐ The global economy will recover sooner rather than later

▨ DK/NR

BASE: Canadians; March 6–11, 2012 (n=2,001).
Copyright 2013. No reproduction without permission.

Source: Graves (2012, 16).

The fraying of the progress ethos surfaces starkly in EKOS's tracking of the public mood. The exuberant optimism in young and old Canadians alike that defined the close of the twentieth century has given way to pessimism and a resignation that steadily has gathered force over the past decade.

The economy is seen to be on a steady downward cycle with many Canadians fearing the spiral will end in a maelstrom. The remedies offered by politicians are viewed as lacking credibility. Trust in government is at a 30-year low (Figure 8). And economic issues have become the dominant concerns for Canadians, twinned for the first time in the history of EKOS research at the pinnacle of public issues with concerns about fairness and inequality (Figure 9). These are not the traditional and more modest concerns we have seen in the past about the gap between rich and poor. This new and more potent linkage is the perceived gap between the über rich and everyone else, and nowhere is this dynamic more evident than in what can only be described as the crisis of the middle class.

Figure 7: No New Taxes versus Taxing the Rich

*Q. In the next federal election, would you be more likely to support a party that
promised to NOT raise taxes or a party that promised to raise taxes on the rich?*

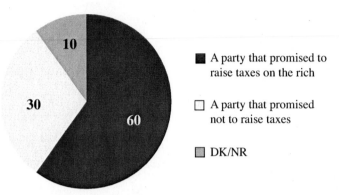

■ A party that promised to
raise taxes on the rich

☐ A party that promised
not to raise taxes

▨ DK/NR

BASE: Canadians; February 21–28, 2012 (n=3,699).

Source: EKOS Research Associates (2012a, 5).

Figure 8: Trust in Government

*Q. How much do you trust the government in Ottawa/Washington to do what is
right?*

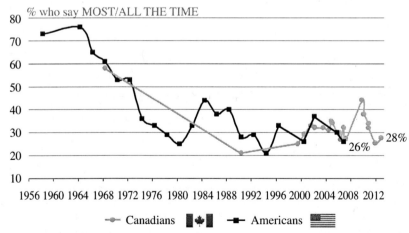

BASE: Canadians; most recent data point November 20–29, 2013 (n=1,500).

Source: EKOS Research Associates (2013b, 18).

Figure 9: Most Important Election Issue

Q. Of the following issues, which one do you think should be the most important issue for the next federal election: 1) issues like ethics and accountability; 2) the economy, jobs, and growth; 3) fiscal issues like taxes and debt; or 4) social issues like health and education?

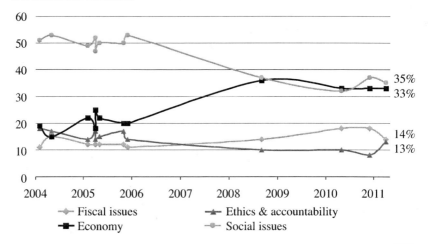

BASE: Canadians; most recent data point April 4–7, 2011 (n=2,204).
Copyright 2013. No reproduction without permission.

Source: EKOS Research Associates (2011, 7).

The middle class has always been by far the most popular self-defined class in upper North America. The twentieth-century ascension of the United States to the "hyper power" status it enjoyed as little as a decade ago was largely the culmination of an unprecedented period of middle-class ascendance that probably began in the origins of that nation, but most clearly expressed itself in the expansionary period following the Great Depression and continued almost uninterrupted until the close of the twentieth century. Canada largely followed its neighbour in lockstep and it was not unusual in the sixties and seventies to see Canada and the United States at the top of the standard-of-living charts. They are now well down that list and have been so for some time; Canada, for example, held the No. 1 position on the United Nations Human Development Index for more than a decade in the 1980s–1990s but was No. 11 in the report released in March 2013 (Ditchburn 2013).

There are few if any modern examples of economic and societal success that have not been defined by a rising, optimistic, and growing middle class. This feature is common to all of the contemporary emerging Asian powerhouses. In contrast, Canada has a shrinking, stagnant, and pessimistic middle class that has lost faith in the ethic of progress (Figure 10). Uncorrected, this will lead to inevitable further decline. The fact that only 14 percent of Canadians think their children will inherit a better world underlines the gloominess of the national mood.

Figure 10: End of Progress

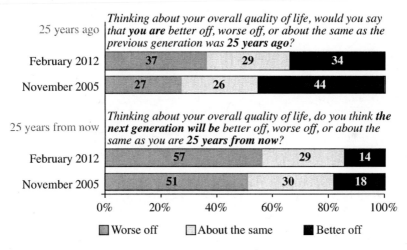

BASE: Canadians (half-sample each); February 21–28, 2012 (n=3,699).
Copyright 2013. No reproduction without permission.

Source: Graves (2013b, 13).

EKOS's tracking over the past decade or so has detected something new and important happening to the category of self-defined middle class: the two out of three Canadians who called themselves middle class has now dropped to about one out of two (Figure 11). Over the past generation, no class has fallen more steeply from economic grace. In sum, while still the largest class label for Canadians, the middle class has shrunk significantly over the past decade. Its members no longer see themselves vaulting in flocks into the parlours of the wealthy. Rather, overall movements are downward, with erstwhile members of the middle class descending into the working and poor classes. Indeed, middle-class Canadians starkly see themselves as the collateral damage of the last quarter-century: When asked to assess the losers and winners of that period, they declare that the rich have gotten richer and everyone else has not (Figure 12), a perception that seems to be feeding broad receptiveness to the idea that less government and lower taxes should mean a return ticket to greater prosperity.

Except that's not happening, which brings us next to the subject of values.

Figure 11: Self-Rated Social Class

*Q. Would you describe you and your household as poor, working class, middle
class, or upper class?*

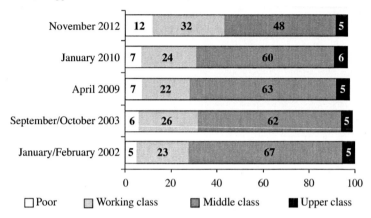

BASE: Canadians; most recent data point November 20–December 3, 2012 (n=5,433).
Copyright 2013. No reproduction without permission.

Source: EKOS Research Associates (2013b, 5).

Figure 12: Long-Term Winners and Losers

*Q. Do you believe the following groups have moved ahead, fallen behind, or
stayed the same over the last 25 years?*

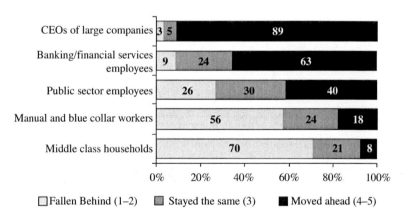

BASE: Canadians; November 20–29, 2012 (n=1,181).
Copyright 2013. No reproduction without permission.

Source: EKOS Research Associates (2013b, 5).

CANADIAN VALUES: GOVERNMENT GOING ONE WAY, THE PUBLIC THE OTHER

Values are propositions about what constitutes good and bad, right and wrong. They are at the heart of what kind of society we want to live in and hand off to the next generation and how we want to be seen by the external world. When values do shift – and they move at a glacial pace – it is extremely important. Unlike more mercurial opinions and attitudes, values constitute a moral charter that underpins societal trajectory. Their importance to national governments is obvious.

They are also tricky things, for as sociologist Michael Mann (1970) has pointed out, while some values unite people, others divide them, depending upon how a particular value is defined. The more consensus there is around the essentialness of a given value, the greater the potential for conflict (for example, conservatives and liberals alike appeal to values of "social justice," "democracy," "peace" and even something that falls under the rubric of "Christian values," and yet the differences in ideological meanings invested in the terms can be great). As Mann writes, "Most general values, norms and social beliefs usually mentioned as integrating societies are extremely vague and can be used to legitimate any social structure, existing or not" (424). Thus there can be no overall common values framework in our increasingly pluralistic world, but it is important to understand the points of consensus and contrariety.

Are values shifting in Canada? The answer is yes, but not, we emphatically point out, in the manner or direction being suggested in mainstream media and by governments, especially Ottawa's. There are, in fact, some huge gaps and distortions *à la* Mann in the popular understanding of national values and how they have been changing.

The most powerful predictor of one's values is one's self-identified political ideology, and many values are grouped clearly into conservative and liberal sets (whereas others, such as freedom and respect and others mentioned above, transcend ideological boundaries). With that in mind, we note some very important changes in ideological orientation that mirror shifts in individual values.

First, Canadians have become much more ideologically polarized in recent years, displacing a previous trend toward de-ideologicalization. For many years after Daniel Bell (1961) wrote of the end of ideology, Canadians seemed to be eschewing ideology in favour of a more pragmatic, eclectic, and politically promiscuous outlook. But the arrival and continued success of Canada's first ever government that clearly (at least rhetorically) governs from the right has disrupted this pattern and produced a newly polarized citizenry that has vacated the centre for havens around either pole. We see this polarization pervasively expressed in more specific attitudes to most of the key policy issues of the day.

The second key change may be even more important than the first, although it is not evident in the media or in popular discussion of shifts in values and ideology.

There have been several recent pronouncements about how Canada is bluing or shifting to the right, a phenomenon that would seem to make sense in light of what is going on in the political marketplace, particularly at the federal level, and to offer legitimacy to further movements in national policy in that direction.

To illustrate, Jason Kenney, at the time minister of immigration and citizenship, told the Manning Networking Conference in March 2013 that the Conservative Party of Canada's share of the popular vote would grow in the next election: "I believe that our party's priorities are closer to those of Canadians. Our values are closer to those of Canadians than any other party" (Pavlich 2013). He identified those values as

> hard work and personal responsibility, a respect for tradition, a belief in family as the most important social institution, a respect for religious faith, a respect for law and order that ensures safe streets and values victims' rights over those of criminals, a belief in entrepreneurship and initiative and risk-taking, the freedom to take chances and reap the rewards of initiative free of crippling taxes and red tape, a belief in an opportunity to succeed by playing by the rules without the over-developed liberal sympathy for those who refuse to do so, belief in a principled, democratic foreign policy that stands up for freedom and fundamental values and a pride in our Canadian armed forces and our history of military sacrifice and glory.

According to Figures 13 and 14, many Canadians appear to share Mr. Kenney's assessment. Indeed, more than twice as many Canadians think we are moving toward the right than toward the left. In a further bit of irony, the perception of a rightward tilt is most pronounced where it is least welcome. Those who lean left are most likely to see a rightward tilt, and the corollary is true for those on the right who see a leftward movement.

Kenny's assertion that the Conservative Party embodies the values of Canadians, however, is at odds with both EKOS's tracking of ideology and its rigorous tracking of core social values. The idea, to begin with, that there can be consensual values framing a pluralistic society such as Canada's is a chimera. Many contradictory values are held tenaciously, which leave little room for central terrain (e.g., right to life/right to choice, capital punishment/abolition, gun control/right to bear arms); many core values are not divisive ideologically (e.g., freedom, respect); and most Canadians hold positive views of both small-c conservative values and small-l liberal values. Nevertheless, Canadians are significantly less connected to socially conservative values than they were 20 years ago, a theme we will return to in more detail below.

Suffice it to say that, looking at values overall, we are struck by their level of stability – something to be expected and welcomed since values constitute the moral charter for societies and it would be a very bewildering and unstable world if they were to shift rapidly. Yet within this placid world of normative stability in Canada, there are some conspicuous exceptions: specifically, all of the values that are demonstrating downward trends are conservative values (Figure 15).

Figure 13: Perceived Shift in Canadian Ideology

Q. When it comes to political ideology, some people say that Canada is moving to the right while others say it is moving to the left. Do you believe Canada is moving right, moving left, or not moving at all?

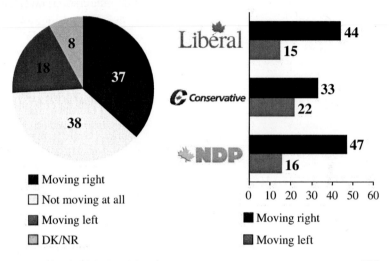

- ■ Moving right
- ☐ Not moving at all
- ▨ Moving left
- ▧ DK/NR

- ■ Moving right
- ▨ Moving left

BASE: Canadians; May 22–26, 2013 (n=3,318).
Copyright 2013. No reproduction without permission.

Note: This study was conducted using Interactive Voice Response (IVR) technology, which allows respondents to enter their preferences by punching the keypad on their phone, rather than telling them to an operator. The field dates for this survey are May 22–26, 2013. In total, a random sample of 3,318 Canadian adults aged 18 and over responded to the survey. The margin of error associated with the total sample is +/-1.7 percentage points, 19 times out of 20.

Source: EKOS internal survey.

Respect for authority and traditional family values, still very important in older and conservative Canada, holds no resonance in younger and university-educated Canada. There are similar downward trends for conservative values such as minimal government (Figure 16) and heightened security – particularly in younger Canada, metropolitan Canada, university-educated Canada, and among women. EKOS research finds a growing appetite for larger (albeit more effective) government. It finds growing skepticism and fatigue with the neo-liberal mantra that less government plus lower taxes equals prosperity for all. It finds declining acceptance for the proposition that tax is a four-letter word.

Figure 14: Political Ideology

Q. Do you consider yourself a small "c" conservative or a small "l" liberal?

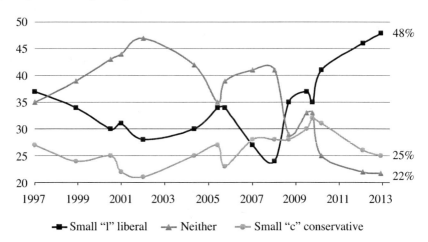

BASE: Canadians; most recent data point November 20–29, 2012 (n=1,181).
Copyright 2013. No reproduction without permission.

Source: Graves (2013b, 7).

We state clearly that there is virtually no plausible evidence to suggest that on social values Canada is shifting to the right. The Conservative Party is doing better politically among immigrants but that does not equate to saying immigrants are moving Canadian values to the right. The success of political parties of the right is not a product of a rightward shift, nor is the presence of a right-of-centre party in Ottawa moving the public to the right. The best indicator of who votes Conservative nationally is religiosity and, in reality, the factors that are moving values are far deeper and transnational than those within the purview of national governments. The value shifts that we see continuing in Canada are part of broader rhythms that are evident throughout the advanced western world (and may be becoming more global in nature).

Thus, while explicitly excluding fiscal conservatism from this claim, we can say without hesitation that the evidence is clear that Canadians are significantly less connected to socially conservative values than they were 20 years ago. Even more important, these values are much less relevant in certain portions of Canadian society such as younger Canada, metropolitan Canada, university-educated Canada, and Quebec. In short, these socially conservative values have little relevance to the emerging, next Canada. While they are highly motivating

Figure 15: Most Important Goals and Values

Q. If you were to direct Canadian society as to which goals or values should be most important in its direction, how important would you say each of the following goals and values should be?

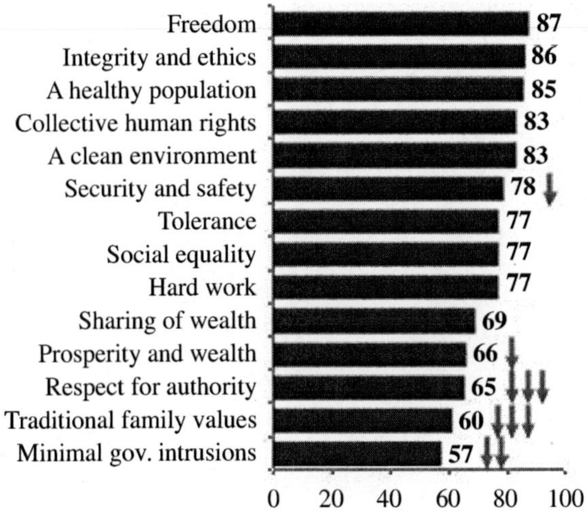

	2005	2002	1998	1994
Freedom	87	90	88	89	89
Integrity and ethics	86	86	84	81	84
A healthy population	85	87	86	86	86
Collective human rights	83	84	80	80	80
A clean environment	83	86	87	83	87
Security and safety	78 ↓	84	82	80	83
Tolerance	77	80	74	75	74
Social equality	77	79	--	77	76
Hard work	77	80	81	79	--
Sharing of wealth	69	74	70	72	70
Prosperity and wealth	66 ↓	68	73	72	73
Respect for authority	65 ↓↓↓	78	81	79	81
Traditional family values	60 ↓↓↓	77	78	79	78
Minimal gov. intrusions	57 ↓↓	65	67	66	67

0   20   40   60   80   100

BASE: Canadians; November 20–29, 2012 (n=1,181).
Copyright 2013. No reproduction without permission.

Source: EKOS Research Associates (2013b, 15).

to the older, core Conservative vote, they are next to meaningless to the groups mentioned above.

All of which lead to obvious questions. If Canada is abandoning (relatively) conservative values, then why is a politically successful conservative federal government in power? What is the relationship between a politically successful conservative government and EKOS's finding that Canadians' confidence in the national direction of the country is nearing a historic low and the temporary surge in trust the federal government enjoyed in the middle of the last decade has disappeared? And what are the implications for these at least ironic cross-current trajectories for the future of national identity and the social fabric, the latter rent by divisions between young and old Canada, urban and rural/small town Canada, and well-educated and less well-educated Canada?

We will address those questions further on. First, we'll deal with the fault lines just mentioned. They are deep and they manifest relatively incompatible world views.

Figure 16: Preferred Size of Government

Q. Generally speaking, which of the following would you say that you favour?

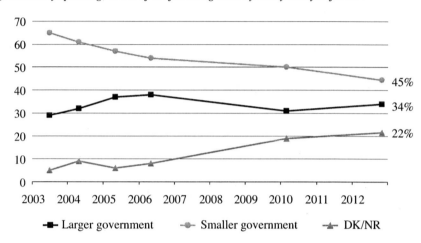

── Larger government   ─●─ Smaller government   ─▲─ DK/NR

Source: Graves (2013b, 36).

THE DEEP FEARS OF THE MIDDLE CLASS

EKOS finds growing conflict between a conservative gerontocracy and a progressive "next" Canada – a "next" Canada seen to be disengaging from formal political participation and with a rising conviction that public institutions favour the old. There is, as noted earlier, a striking left-right split linked to educational attainment. And the relative salience of reason and knowledge versus moralism and certainty is contested terrain, with the rational-empirical view much more prevalent among younger and better-educated Canadians.

Aggravating these fractures are economic fears, fears of a generational decline, fears of an eroding middle class, fears that inequality is removing the middle rungs of the economic ladder, a growing resentment of the upper classes (unfamiliar language to describe Canadian society), fears of the eroding relative global positions of the European and US/Canadian economies, and a steady rise in pessimism for the future of the progress ethos – the belief in inevitable social and economic betterment – that has been the underpinning of the Canadian economy for at least as long as there has been a Canada. In sum, Canada has moved from the 9/11-inspired need-for-security decade (Figure 17) to the economic anxiety decade (Figures 18 and 19).

Figure 17: Additional Powers for Law Enforcement

Q. To what extent do you agree or disagree with the following statement: police and intelligence agencies should have more powers to ensure security even if it means Canadians have to give up some personal privacy safeguards?

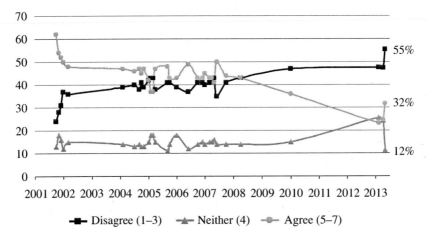

BASE: Canadians; most recent data point April 30–May 2, 2013 (n=1,309).
Copyright 2013. No reproduction without permission.

Source: Graves (2013a).

It presents a toxic combination: economic insecurity plus fractures over age, education, class, as well as urban/rural domicile and even gender. Moreover, at the root of Canadians' – primarily middle-class Canadians' – economic anxiety is the spectre of inequality. What kind of national society allows these discontents and disconnections and miseries to befall its citizens? The implications for public life in common and social cohesion and identity are not good.

Discussions of class structure and class tensions tend not to be top-of-the-head topics of conversation for Canadians. We are inclined to think of ourselves, however imperfect the notion, as a relatively classless society. Yet concerns with the "middle class" have now become a mainstay of politicians in Canada and the United States, similar to how "family" only a few years back was the pitch-word programmed for political triumph. Beyond the bromides about the importance of a healthy middle class and about how middle-class people need to find their lives affordable and optimistic again, there are truly profound changes happening that are decidedly unhealthy and, taken together, may spell a deep rupture in the castle-in-the-air of perpetual progress that has underpinned liberal capitalism since it emerged in the eighteenth century.

Figure 18: Long-Term Personal Financial Outlook

*Q. Thinking ahead over the next FIVE YEARS or so, do you think your personal
financial situation will be better or worse than it is today?*

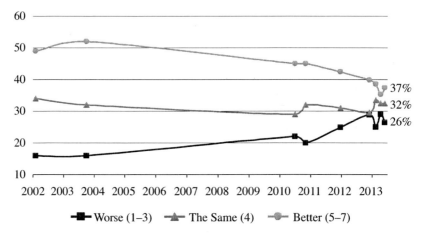

BASE: Canadians; most recent data point May 22–26, 2013 (n=3,318).
Copyright 2013. No reproduction without permission.

Note: This study was conducted using Interactive Voice Response (IVR) technology,
which allows respondents to enter their preferences by punching the keypad on their
phone, rather than telling them to an operator. The field dates for this survey are May
22–26, 2013. In total, a random sample of 3,318 Canadian adults aged 18 and over
responded to the survey. The margin of error associated with the total sample is +/-1.7
percentage points, 19 times out of 20.

Source: Internal EKOS survey.

Despite pronouncements about the end of history, the death of state socialism,
and the final triumph of free-market capitalism, all pervasive themes in the late
nineties, there are now grave doubts about the present and future of the advanced
western economies. The American and Canadian dreams of a better future extracted
from hard work and ingenuity are fading and being replaced with a grimmer sense
that not only are we not doing better than our parents but the next generation will
confront a starkly darker future, and what meagre profits do emanate from stagnant
western economies are increasingly appropriated by a tiny cadre of über rich who
don't really participate in the mainstream of society.

Charles Beach (2013), Queen's University emeritus professor of economics, writes:

> While the share of income of the poorest 20 percent of families has remained roughly
> the same since the late 1970s, between 1977 and 2010 the share of income of the
> middle 60 percent of families fell from 56.1 percent to 50.7 percent, or by about the

same amount as the income share of the top 20 percent has gone up. So the rising top income share has come at the expense of a falling middle class income share.

[In addition,] there has been a decline in economic mobility in Canada, resulting in receding opportunity to get ahead. Between the 1980s and the 1990s, the average probability of moving up or down in earnings classes over an eight-year period for male earners fell from 64.7 percent to 62.7 percent, and it fell for female earners from 59.9 percent to 58.4 percent. The probability of moving up across earnings classes also fell for both men and women. Evidence also suggests that it is getting harder for sons to move up the economic ladder than their fathers' generation. (20)

Beach concludes: "We may be facing a historic shift away from a period in which middle class Canadians could expect their economic prospects to be better than their parents' were and more of them could expect to move up the economic ladder. Given that the middle class is where most Canadian votes reside, their reaction may carry considerable political consequences. While rising incomes at the top end of the income distribution have received most of the media attention, it is the declining middle income shares that will likely be the more politically potent concern" (20).

Figure 19: Future Well-Being of World Economic Zones

Q. Over the next decade, the world economy will undergo many significant changes. Do you believe that the following economic zones will be better off or worse off ten years from now than they are today?

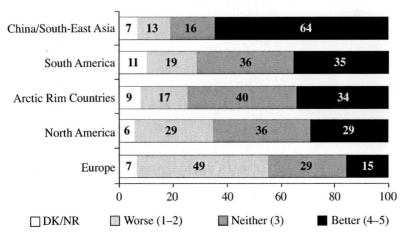

BASE: Canadians; March 6–11, 2012 (n=2,001).
Copyright 2013. No reproduction without permission.

Source: Graves (2012, 17).

In the United States, it has been recognized that institutional precepts and investments favour the rich. In Canada, we haven't gone as far down that road because our social institutions were put in place at a time of low inequality. But the warning signs are undeniably present. Inequality is real. Our labour market is becoming much more polarized. Our middle class faces higher and higher barriers to investing in their own future. Those are facts of which Canadian workers are starkly aware in the face of rather relentless efforts by Canada's mainstream media, right-wing think-tanks, and financial institutions to demonstrate otherwise. On its inequality scale, the Conference Board of Canada ranks Canada 12th of 17 peer countries – meaning income inequality is higher in Canada than in 11 of its peers – as measured by the standard Gini coefficient (Lafleur et al. 2013). Statistics Canada reports that the median earnings of full-time Canadian workers rose to $41,401 in 2005 from $41,348 in 1980 – an increase of $53 over 25 years or about $1 a week measured in constant dollars. In that period, the incomes of the richest Canadians increased by 16.4 percent while incomes of the poorest fell by 20.6 percent, meaning the incomes of the middle class either stagnated or went down (Valpy 2008). Meanwhile, personal income tax levels and capital gains and corporate tax levels are all dramatically lower today than they were 40 years ago (Cadesky and Weissman 2013), and the proportion of unemployed Canadians receiving jobless benefits is lower today than it was in 1945 (Yalnizyan 2009, 32, updated by author in 2013).

Middle-class erosion has been associated with a weakening union movement and with the disappearance of manufacturing and middle-management jobs that paid a reasonable living. It is very much being aggravated by Canada's rapid deindustrialization as well as by globalization, by the offshore exporting of jobs, and by technological change which, granted, no government has the power to halt. It is affecting our multiculturalism, the thing Canadians have done best at in recent decades and which now shows signs of beginning to bore us at a time when Canadians should be examining it closely to see whether it is falling among thorns. Opposition to immigration is less than half what it is in the United States. EKOS (2006) reported that 20 years ago, in 1994, roughly half of Canadians felt there were too many immigrants; by 2006, this figure had fallen to just 27 percent. Today it would be surprising if that number reached into double digits, although current economic anxieties show signs of driving enthusiasm for immigration downward in the long term (EKOS 2013a).

Multiculturalism may be the reason that Canada has become the world's most successful postmodern and postracial society. Our mantra of unity through diversity – mosaic not melting pot – may be why Canada is so relatively immune to the so-called clash of civilizations that has been proclaimed in Europe and America. It should be the Canadian advantage. And yet 60 percent of immigrants to Canada – primarily people of colour – fail to find jobs commensurate with their education and training. Ninety-five percent of immigrant doctors never qualify to practice their profession in Canada. Statistics Canada reported in 2008 that university-educated immigrants earn less than half what university-educated native-born Canadians earn in comparable jobs. Psychiatrist Kwame McKenzie, a senior scientist at Toronto's Centre for Addiction and Mental Health, says that immigrants, when they arrive in Canada, are healthier on average than native-born Canadians; seven years later they are unhealthier, primarily because of the stresses that envelope them as they try to fit into Canada's labour force and culture (Goar 2013, A11; Valpy 2013). It is not just citizens being squeezed out of the middle class; it is newcomers not being allowed into the middle class.

And along with everything else – and here we come to the real meaning of inequality, the indicator of equity in the economy – middle-class Canadians are beginning to perceive their institutions as being bent to favour the rich: for example, more pressure on governments to allow private-sector health care for the wealthy who don't like waiting in line, and more foreign caregivers being admitted for the children and elderly relatives of the well-to-do. Meanwhile, universal childcare and expanded public nursing care programs have fallen off the political agenda.

Perhaps the reason why all this is not recognized as a crisis is because it's a slow downward grind rather than a catastrophic shock, and therefore many Canadians are inured to what is happening, the school of thank-God-we're-not-Greece-so-let's-carry-on. EKOS has found that one of the interesting features of the Conservative Party support base is that it blends those who like small-c conservative values with those still optimistic about their economic futures. Thus both values and economic self-interest unify the Conservative Party base in a way that does not unify or motivate those who worry about their economic futures or do not connect with social conservative values.

Inequality is more than a moral issue relegated simply to a trade-off with efficiency. The classic argument (as the Conference Board puts it) is "that more income equality reduced incentives to work harder, to invest, and to get more education" (Lafleur et al. 2013, "Income Inequality"). However, current thinking on the issue has evolved and the Conference Board quotes a 2011 study by the International Monetary Fund that found the following:

> When growth is looked at over the long term, the trade-off between efficiency and equality may not exist. In fact equality appears to be an important ingredient in promoting and sustaining growth. The difference between countries that can sustain rapid growth for many years or even decades and the many others that see growth spurts fade quickly may be the level of inequality. Countries may find that improving equality may also improve efficiency, understood as more sustainable long-run growth. (Berg and Ostry 2011, 13)

In fact, the IMF authors go on to argue that the recent global financial crisis "may have resulted, in part at least, from the increase in inequality" (*Economist*, October 13, 2012, 9). This is because, they say, inequality tends to be related to financial crises – as inequality rises, people on the bottom of the income scale tend to borrow more in order to keep up, which in turn increases the risk of a major crisis. Second, severe inequality increases social and, in turn, political instability, which reduces foreign investment (Lafleur et al. 2013, "Income Inequality"). And, indeed, on that final point, the *Financial Times* reported in 2010 that "some of the brightest minds at Moody's rating agency have been mulling a fascinating question." Should they introduce a formal rating of "social cohesion" into sovereign debt indices, when they judge whether a government is likely to default on its debt or not? "So far," said the *Financial Times*, "neither Moody's nor any other agency has actually done this. But the discussion points to a fundamental issue that will hang over bond markets this decade" (Tett 2010, 20). In fact, a Canadian academic expert on the global rating agencies, speaking on condition of confidentiality because of the nature of his research work, has told the authors of this chapter that Moody's does indeed use a social cohesion indice in determining sovereign debt ratings (interview, July 10, 2013).

AGE FRACTURE, SOCIAL MEDIA, AND THE EMERGENT GERONTOCRACY

We turn now to the greying of Canada and why growing youth disengagement from the country's formal democracy could not be more poorly timed against this demographic backdrop.

Canadian society has never been older. The more apocalyptical grey tsunami-to-come scenarios are no doubt exaggerated, yet there is something disturbing about the new generational fault lines in both the economic and, even more vividly, the political realms.

Beginning first with youth unemployment, the Conference Board of Canada (Lafleur et al. 2013) ranks Canada no better than ninth out of 16 peer countries, with joblessness among young people (ages 20–24 not in employment, education, or training [NEET]) at almost 14 percent, double the national average. Youth unemployment has not budged since the 2008 recession – the most severe recession since the 1930s – and virtually no federal government programs have been created to target youth joblessness. Young Canadians have tried hard to adjust to the labour market that confronts them. They have delayed marriage, delayed children, acquired more and more education. None of it is helping.

The generation Y/millennials labour force entrants – born after 1981, the largest population cohort to come along since the baby boomers – find a job market cluttered at the far end with entrenched boomers immersed in morphing freedom-55 into freedom-75 and beyond. (That being said, the OECD notes that intergenerational solidarity is at its strongest when older people are seen to be taking actions to help themselves, through private saving for retirement and/or continuing to work, and that keeping older workers in the labour force does not reduce job opportunities for the young. "This is clear both from looking at patterns of employment of younger and older workers across countries and in a single country over time." Perceptions may be different [Martin and Whitehouse 2012].)

Canada's Gen Y student cohort finds itself increasingly with only one means – borrowing – to move up the social-class rungs. They are weakening in their belief that post-secondary human capital is worth the ever-mounting debt associated with its achievement (Figure 20). Canadian employers are not giving them the crucial work experience needed to make them appealing for full-time employment, with the result that what skills they do have are in danger of atrophying the longer they go without work. Canada trails most peer countries in spending on active labour market programs such as training and skills development. The Conference Board reports that Canada has not improved its ranking on absorbing young people into the labour market for over three decades (Lafleur et al. 2013). Canada facilitates the training of far fewer skilled-trades workers than the economy requires. In 2010, only 6 percent of upper secondary students were enrolled in vocational or prevocational programs, the lowest rate among peer countries from which data are available; in nine peer countries, more than half of upper secondary students were in vocational

Figure 20: Benefits of a University Education

Q. Please rate the extent to which you agree or disagree with the following statement: "The cost of a university education is a good, long-term investment for today's young people."

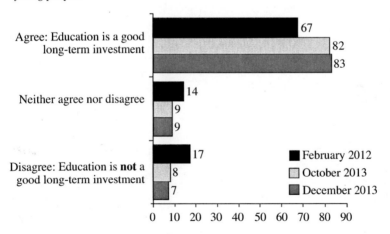

Note: This online-only study was conducted using EKOS's research panel, Prob*it*. The field dates for this survey are July 16–23, 2014. In total, 2,891 Canadians aged 18 and over responded to the survey. The margin of error associated with the total sample is +/-1.8 percentage points, 19 times out of 20.

Source: Internal EKOS survey.

programs (Lafleur et al. 2013). Meanwhile, Canadian businesses press government to allow skilled trades workers into the country who have been trained elsewhere, thus sparing Canada the cost.

Across the Atlantic, there are warning signs on what failing to absorb young people into the labour market can mean. Thirteen of the European Union's 27 member states have youth unemployment above 25 percent. The Brussels-based think-tank Bruegel says youth unemployment could reverse Europe's slow financial recovery, and the Roman Catholic global aid and development agency Caritas, in a February 2013 report, warned that eurozone countries are creating a huge class of poorly educated and poorly fed young people with low morale and few job prospects. "This could be a recipe not just for one lost generation in Europe but for several lost generations," the agency said (quoted in Davenport 2013, A12).

Canada's millennials have such shining promise. They are much more ethnically diverse than older Canada. They grew up digitally. They are the first generation to have more women than men obtain post-secondary credentials. They have different attitudes to community, privacy, and authority than their older fellow citizens. They are much better educated (as well as being more secular) than previous generations. They represent a widening generational gap on core values. When Jason Kenney talks about "mainstream Canadian values" being Conservative values, it is not Gen Y's values he is talking about, but rather the beliefs and convictions of the aging baby boomers, the people heading toward the ends of their journeys through the Canadian workforce.

All of these differences place young and old Canada in conflict, albeit with nuances. Today's generational tensions may be no less than the inflamed tensions of the sixties and early seventies that brought dramatic reforms to racial discrimination and civil rights, women's equality, and the end of the cold war along with demands by young people for a more caring and sensitive Canada toward its marginalized citizens: the poor, the disadvantaged, homosexuals, and racial minorities.

But one does not get the sense that today's Gen Y possesses the same activist heat as its sixties' predecessors (certainly in anglophone Canada; Quebec may be a different story). Indeed, one recent study has found a disturbingly large clump of young Canadians – whom it dubs "The Spectators" – who do not participate in most of the interactivity that society values, show no interest in public affairs have limited life-goals and aspirations, and spend more time online than the average Canadian (Herle 2012, 19), reinforcing the thought that Neil Postman (1985) got it right in his seminal 1985 book titled *Amusing Ourselves to Death.*

The Internet is the new mass media and social media, now avidly consumed by most Canadians, particularly those below the country's median age of 42 (it was around 26 at the Centennial celebrations of 1967). Social media is the news medium of choice for 52 percent of Canadians (Figure 21).This is more than a change to our popular culture: social media is at the heart of the North American economy with the Facebook IPO being acclaimed as the biggest economic event of 2012. Notably, on the day that Facebook purchased Instagram for US$1 billion, the *New York Times* was valued at US$900 million.

Figure 21: Preferred News Medium

*Q. How frequently do you follow **political and governmental affairs** in each of the following?*

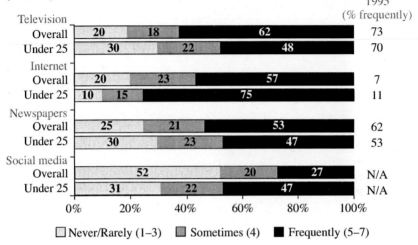

1995
(% frequently)

Television		
Overall	20 / 18 / 62	73
Under 25	30 / 22 / 48	70
Internet		
Overall	20 / 23 / 57	7
Under 25	10 / 15 / 75	11
Newspapers		
Overall	25 / 21 / 53	62
Under 25	30 / 23 / 47	53
Social media		
Overall	52 / 20 / 27	N/A
Under 25	31 / 22 / 47	N/A

0% 20% 40% 60% 80% 100%

☐ Never/Rarely (1–3) ▨ Sometimes (4) ■ Frequently (5–7)

BASE: Canadians; April 30–May 3, 2013 (n=1,309).
Copyright 2013. No reproduction without permission.

Source: Graves (2013a).

When EKOS asked Canadians their views on what impacts social media are having on quality of life in general and democratic health in particular, the responses were overwhelmingly positive. As shown in Figure 22, virtually everyone thinks that social media is a liberating force that is enriching and broadening democratic and societal health. Putting aside the irony that this consensus comes at a time when barometers of democratic health are at historical low points in our tracking, we are left puzzled about this nearly unanimous thumbs-up on the salubrious impacts of social media.

In Canada, there has been an explosion of interest in the use of social media as a form of political expression. Online communities and petitions abound, and the Twitterverse is awash in critical commentary of the most dramatic sort. But with what effect? What impact? Mike Colledge of Ipsos-Reid noted that during the May 2011 federal election campaign the tone of the Internet shifted from a relatively balanced ideological voice to a decidedly more left-of-centre voice (Graves 2013c, 2). The shift had no effect on the election's outcome.

Some have argued (for example, see *Policy Options*, November 2012) that the less strenuous "click democracy" available to inhabitants of the social media universe is becoming an ersatz touchstone that occludes the importance of authentic

Figure 22: Role of Social Media in Democracy

*Q. As you may know, the use of social networking websites such as Facebook and Twitter has increased dramatically in recent years. **Some people argue that social media is good for democracy** since it offers new ways of participating in politics and communicating with the public. **Other people argue that social networking is harmful to democracy**, since many people will use these websites as a substitute for real world action. **Which of these statements comes closest to your own point of view?***

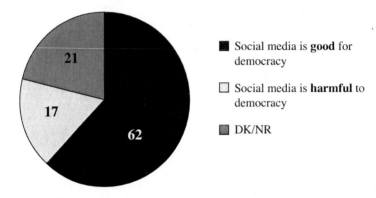

- Social media is **good** for democracy
- Social media is **harmful** to democracy
- DK/NR

BASE: Canadians; December 14–21, 2011 (n=2,005).
Copyright 2013. No reproduction without permission.

Source: EKOS Research Associates (2012b, 20).

political participation. Moreover, those who vigorously contest the policies of the day in the world of social media, and who believe that this is really making a difference, become more embittered as this delusion is shattered in the real world of elections. In Canada, youth voting has not risen in tandem with the rise of social media – to the contrary.

Thus what generational tensions may mean for the future of social cohesion in Canada is so far a mystery; no one is really quite sure what Gen Y's sense of the collective is. But couple their edginess and their divergence of values from those of the government's supportive gerontocracy with the unusually grim outlook on the country's economic future, and we can see the ingredients of a major problem for an aging society that desperately needs the innovation and dynamism of its younger cohort to confront the challenges of economic stagnation. And when we examine general behaviour in the realm of politics in Canada, the scenario we see is frankly alarming.

Twenty years ago, younger and older voters were approximately similarly sized portions of the electorate. Today older voters comprise a relative 50 percent larger share of the overall electorate. Couple that statistic with what has happened to the voter participation rate of younger Canadians. In the 1993 federal election, they participated slightly less than seniors at around 65 percent. Today their rate is about half that (38.8 percent of the population aged 18 to 24 voted in the May 2011 federal election) while senior voting has remained constant. Effectively, a younger voter has about one-third to one-quarter the impact today that she or he did 20 years ago. Throwing one final ingredient into the mixture, we note that while the senior vote tended in years past to be more or less evenly split between the Liberal and Conservative parties, it now has converged dramatically around the Conservatives (Figure 23). Putting these three factors together – an increasing old vote, a declining young vote, and an old vote glued to the Conservatives – goes a long way to explaining why a federal government that champions values of security, safety, respect for authority, family values, and so on has been so successful.

Figure 23: Federal Vote Intention

Q. If a federal election were held tomorrow, which party would you vote for?

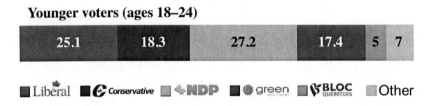

National Results

| 34.7 | 26.3 | 20.9 | 9.2 | 5 | 4 |

Younger voters (ages 18–24)

| 25.1 | 18.3 | 27.2 | 17.4 | 5 | 7 |

■ Liberal ■ Conservative ■ NDP ■ green ■ BLOC QUÉBÉCOIS ■ Other

Note: The data on federal vote intention are based on decided and leaning voters only. Our survey also finds that 17.8% of respondents are undecided and 2.5% are ineligible to vote.

BASE: Canadians; May 22–26, 2013 (n=3,318).
Copyright 2013. No reproduction without permission.

Note: This study was conducted using Interactive Voice Response (IVR) technology, which allows respondents to enter their preferences by punching the keypad on their phone, rather than telling them to an operator. The field dates for this survey are May 22–26, 2013. In total, a random sample of 3,318 Canadian adults aged 18 and over responded to the survey. The margin of error associated with the total sample is +/-1.7 percentage points, 19 times out of 20.

Source: EKOS internal survey.

For the Conservatives, it obviously makes great sense to consolidate support around emotionally resonant policies and communications that will appeal to a group that will vote en masse for them. By corollary, it makes sense to discourage the participation of younger voters (who would not vote for the Conservatives if they were to show up) through negative advertising and policy positions that are of little or reverse interests to younger voters.

The net result is a gerontocracy that reflects the exaggerated and imagined fears of older Canada precisely at a time when the country urgently needs the more optimistic and innovative outlooks of the relatively scarcer younger portion of our society. Where, in all this narrative, are the mechanisms of intergenerational solidarity – the investment in young Canadians and their pursuit of education and jobs in return for their obligations to finance the home care, health care, and pensions for the old? Thus politics becomes highly suspect as a tool for meeting the severe challenges of the twenty-first century. This growing disjuncture between the public interest and what works in the realm of the political marketplace is a stern challenge, and the mounting generational tensions in our society are just one particularly unwelcome expression of this.

SOME REMEDIES

Half a century ago, University of Toronto English scholar Germaine Warkentin wrote:

> Searchers for a Canadian identity have failed to realize that you can only have an identification with something you can see or recognize. You need, if nothing else, an image in a mirror. No other country cares enough about us to give us back an image of ourselves that we can even resent. And apparently we can't do it for ourselves, because so far our attempts to do so have resembled those of the three blind men trying to describe the elephant. Some of the descriptions have been worth something, but what they add up to is fragmented, indecipherable. With what are we to identify ourselves? (1964, 73-74)

Not much apparently – but that's the fate of a people who live at the margins of empire, although Stephen Harper's government is doing something surprising to Canadians: manufacturing resentment toward the country internationally. As for Warkentin's pronouncement on indecipherable fragments, it still carries bleak truth, raising questions about the resilience of our social cohesion, whether there is an "us" that can be talked about, or whether the canescent political science maxim of Canada as a nation of limited identities has taken on new and dismaying muscular meaning.

That, of course, need not be the case. With Canadians' revived interest in the state as an amulet in their collective lives, they might plant a foot on the behind of Parliament with the intent of encouraging its elected members to address issues raised in this chapter.

We cannot push back the tides of technology and globalization, but in the age of the precarious workplace we can look at the Danish system of "flexicurity" and see what Canada can borrow. We can figure out a way of putting a floor under wages so that we are not socially excluding people who cannot afford adequate accommodation, who can't invest in family, who are young and can't afford to live in the cities in which they grew up, who are newcomers to the country and have yet to become fully integrated into the economy.

We can examine our education system to see what recalibrations are required. More than a decade ago, the Dutch turned their school system upside down and placed their major resources into the vocational stream. In nine peer countries, more than half of secondary students are in vocational programs. The 6 percent figure for Canada is self-evidently too low.

We can encourage intergenerational solidarity by widening the pathway for adequately trained young Canadians to enter the workforce and obtain dignified jobs and by collectively guaranteeing dignified support for older Canadians at the end of their active working lives.

Bernard Ostry (2005), the author of Pierre Trudeau's multiculturalism policy, wrote just before his death in 2005 that multiculturalism was coming under increasing criticism and that Canadians no longer were sure where it was taking them. He recommended the country do what it had so successfully done in the past when confronted with major concerns such as health care, broadcasting, immigration, federal-provincial relations, and culture – establish a royal commission to replace darkness with the light of understanding. It would be worse than tragedy if Canada were to abandon one of its finest accomplishments merely to serve short-term ideological or economic ends.

And, finally, voting studies have repeatedly shown that if young people do not develop the practice of voting early in their adult lives, they stay out of the electorate forever. The indications are very strong that this is happening, and for the sake of our democracy, it should not continue unchecked. The time has come for a trial run with mandatory voting.

REFERENCES

Beach, Charles M. 2013. "When the Middle Class Meets Slower Growth." *Policy Options* (March).

Bell, Daniel. 1961. *The End of Ideology: On the Exhaustion of Political Ideas in the Fifties.* new rev. ed. New York: Collier Books.

Berg, Andrew G., and Jonathan D. Ostry. 2011. "Equality and Efficiency." *Finance and Development* 48 (3). International Monetary Fund. http://www.imf.org/external/pubs/ft/fandd/2011/09/pdf/berg.pdf.

Cadesky, Michael, and Peter Weissman. 2013. "Income Tax, Then and Now." *CAmagazine*, June-July.

Davenport, Claire. 2013. "Europe Warned of Generations 'Lost' to Poverty." *Toronto Star*, February 15.

Ditchburn, Jennifer. 2013. "Canada No Longer One of Top 10 Most Developed Countries: United Nations." *Canadian Press*, March 14.

EKOS Research Associates. 2006. "Rethinking Government 2006 Wave 3 Report." Ottawa, December.

—. 2011. "Week Two: Ballot Question Clarifying, Outcome Not." Ottawa, April 8. http://www.ekospolitics.com/wp-content/uploads/full_report_april_8_2011x.pdf.

—. 2012a. "A Divided Public Poses Deep Budget Challenges." Ottawa, March 5. http://www.ekospolitics.com/wp-content/uploads/full_report_march_5_2012.pdf.

—. 2012b. "Beyond the Horserace." Ottawa, January 14. http://www.ekospolitics.com/wp-content/uploads/beyond_the_horserace_jan_14_2012.pdf.

—. 2013a. "Choosing a Better Future?" Ottawa, July 26. http://www.ekospolitics.com/wp-content/uploads/full_report_july_26_2013.pdf.

—. 2013b. "Looking Backward, Looking Forward." Ottawa, January 9. http://www.ekospolitics.com/wp-content/uploads/looking_backward_complete_series_january_9_2013.pdf.

—. 2013c. "Social Media, Socioeconomic Status, and Democratic Health." Ottawa, January 4. http://www.ekospolitics.com/wp-content/uploads/addendum_january_4_2013.pdf.

Goar, Carol. 2013. "Canada Needs a Better Way to Select Immigrants." *Toronto Star*, February 11.

Graves, Frank. 2012. "An Increasingly Divided Outlook: Rethinking Canada's Place in the World." Presentation to the 2012 Walter Gordon Symposium in Public Policy, Toronto, March 20. http://www.ekospolitics.com/wp-content/uploads/2012_walter_gordon_symposium_presentation.pdf.

—. 2013a. "How Changing Values, Demographics, and Economic Context Are Reshaping Canada." Presentation to the 2013 INK+BEYOND Conference, Ottawa, May 3.

—. 2013b. "Left-Right? Forward-Backward? Examining Longer Term Shifts in Values, Social Class, and Societal Outlook." Presentation to the School of Public Policy and Governance at the University of Toronto, March 21. http://www.ekospolitics.com/wp-content/uploads/left_right_forward_backward_march_21_2013.pdf.

—. 2013c. "Looking Backward, Looking Forward: Part 3." *EKOS Politics/iPolitics.ca*, January 3. http://www.ekospolitics.com/wp-content/uploads/looking_backward_part_3_january_3_2013.pdf.

Graves, Frank, with Patrick Beauchamp and Tim Dugas. 1999. "Identity and National Attachments in Contemporary Canada." In *Canada: The State of the Federation 1998/99: How Canadians Connect*. Montreal: McGill-Queen's University Press and School of Policy Studies, Queen's University.

Gwyn, Richard. 1995. *Nationalism Without Walls: The Unbearable Lightness of Being Canada*. Toronto: McClelland and Stewart.

Herle, David. 2012. "Clicktivism – The Spectators." *Policy Options* (November).

Hutchison, Bruce. 1948. *The Unknown Country: Canada and Her People*. rev. ed. Toronto: Longmans, Green.

Ignatieff, Michael. 2009. *True Patriot Love: Four Generations in Search of Canada*. Toronto: Viking Canada.

Jeannotte, Sharon, Dick Stanley, Ravi Pendakur, Bruce Jamieson, Maureen Williams, and Amanda Aizlewood. 2002. "Buying In or Dropping Out: The Public Policy Implications of Social Cohesion Research." Strategic Research and Analysis, Strategic Planning and Policy Coordination, Department of Canadian Heritage. January.

Lafleur, Brenda et al. 2013. "How Canada Performs: A Report Card on Canada." Conference Board of Canada.

Mann, Michael. 1970. "The Social Cohesion of Liberal Democracy." *American Sociological Review* 35 (June): 423-39.

Martin, John, and Edward Whitehouse. 2012. "The Conflict between Generations – Factor or Fiction?" *OECD Observer*. No 290-291, Q1-Q2 2012.

Ostry, Bernard. 2005. "Digging Up Identity Issues." *Globe and Mail*, November 15, A27.

Pavlich, Alexis (press secretary to the Hon. Jason Kenney). 2013. Unofficial transcript of Minister Kenney's remarks to the Manning Networking Conference, Ottawa, March.

Postman, Neil. 1985. *Amusing Ourselves to Death: Public Discourse in the Age of Show Business*. 20th anniversary ed. New York. Penguin Books.

Tett, Gillian. 2010. "Future Funding Strategies Could Prove a Test of Patriotism." *Financial Times*, January 8, Asia Ed1.

Valpy, Michael. 2008. "The Canadian Dream? 25 Years: 53 Bucks." *Globe and Mail*, May 2.

—. 2013. "Does Good Policy Make Good Neighbours?" *Literary Review of Canada* (April).

Warkentin, Germaine. 1964. "An Image in a Mirror." *Alphabet#8* (June). London, Ontario.

Yalnizyan, Armine. 2009. "EXPOSED: Revealing Truths about Canada's Recession." Canadian Centre for Policy Alternatives, Ottawa. Updated by author in 2013.

Queen's Policy Studies
Recent Publications

The Queen's Policy Studies Series is dedicated to the exploration of major public policy issues that confront governments and society in Canada and other nations.

Manuscript submission. We are pleased to consider new book proposals and manuscripts. Preliminary inquiries are welcome. A subvention is normally required for the publication of an academic book. Please direct questions or proposals to the Publications Unit by email at spspress@queensu.ca, or visit our website at: www.queensu.ca/sps/books, or contact us by phone at (613) 533-2192.

Our books are available from good bookstores everywhere, including the Queen's University bookstore (http://www.campusbookstore.com/). McGill-Queen's University Press is the exclusive world representative and distributor of books in the series. A full catalogue and ordering information may be found on their web site **(http://mqup.mcgill.ca/)**.

For more information about new and backlist titles from Queen's Policy Studies, visit http://www.queensu.ca/sps/books.

School of Policy Studies

Toward a Healthcare Strategy for Canadians, A. Scott Carson, Jeffrey Dixon, and Kim Richard Nossal (eds.) 2015. ISBN 978-1-55339-439-6

Work in a Warming World, Carla Lipsig-Mummé and Stephen McBride (eds.) 2015. ISBN 978-1-55339-432-7

Lord Beaconsfield and Sir John A. Macdonald: A Political and Personal Parallel, Michel W. Pharand (ed.) 2015. ISBN 978-1-55339-438-9

Canadian Public-Sector Financial Management, Second Edition, Andrew Graham 2014. ISBN 978-1-55339-426-6

The Multiculturalism Question: Debating Identity in 21st-Century Canada, Jack Jedwab (ed.) 2014. ISBN 978-1-55339-422-8

Government-Nonprofit Relations in Times of Recession, Rachel Laforest (ed.) 2013. ISBN 978-1-55339-327-6

Intellectual Disabilities and *Dual Diagnosis: An Interprofessional Clinical Guide for Healthcare Providers*, Bruce D. McCreary and Jessica Jones (eds.) 2013. ISBN 978-1-55339-331-3

Rethinking Higher Education: Participation, Research, and Differentiation, George Fallis 2013. ISBN 978-1-55339-333-7

Making Policy in Turbulent Times: Challenges and Prospects for Higher Education, Paul Axelrod, Roopa Desai Trilokekar, Theresa Shanahan, and Richard Wellen (eds.) 2013. ISBN 978-1-55339-332-0

Building More Effective Labour-Management Relationships, Richard P. Chaykowski and Robert S. Hickey (eds.) 2013. ISBN 978-1-55339-306-1

Navigationg on the Titanic: Economic Growth, Energy, and the Failure of Governance, Bryne Purchase 2013. ISBN 978-1-55339-330-6

Measuring the Value of a Postsecondary Education, Ken Norrie and Mary Catharine Lennon (eds.) 2013. ISBN 978-1-55339-325-2

Immigration, Integration, and Inclusion in Ontario Cities, Caroline Andrew, John Biles, Meyer Burstein, Victoria M. Esses, and Erin Tolley (eds.) 2012. ISBN 978-1-55339-292-7

Diverse Nations, Diverse Responses: Approaches to Social Cohesion in Immigrant Societies, Paul Spoonley and Erin Tolley (eds.) 2012. ISBN 978-1-55339-309-2

Making EI Work: Research from the Mowat Centre Employment Insurance Task Force, Keith Banting and Jon Medow (eds.) 2012. ISBN 978-1-55339-323-8

Managing Immigration and Diversity in Canada: A Transatlantic Dialogue in the New Age of Migration, Dan Rodríguez-García (ed.) 2012. ISBN 978-1-55339-289-7

International Perspectives: Integration and Inclusion, James Frideres and John Biles (eds.) 2012. ISBN 978-1-55339-317-7

Dynamic Negotiations: Teacher Labour Relations in Canadian Elementary and Secondary Education, Sara Slinn and Arthur Sweetman (eds.) 2012. ISBN 978-1-55339-304-7

Where to from Here? Keeping Medicare Sustainable, Stephen Duckett 2012. ISBN 978-1-55339-318-4

International Migration in Uncertain Times, John Nieuwenhuysen, Howard Duncan, and Stine Neerup (eds.) 2012. ISBN 978-1-55339-308-5

Centre for International and Defence Policy

Afghanistan in the Balance: Counterinsurgency, Comprehensive Approach, and Political Order, Hans-Georg Ehrhart, Sven Bernhard Gareis, and Charles Pentland (eds.), 2012. ISBN 978-1-55339-353-5

Institute of Intergovernmental Relations

Canada: The State of the Federation 2011, Nadia Verrelli (ed.), 2014. ISBN 978-1-55339-207-1

Canada and the Crown: Essays on Constitutional Monarchy, D. Michael Jackson and Philippe Lagassé (eds.), 2013. ISBN 978-1-55339-204-0

Paradigm Freeze: Why It Is So Hard to Reform Health-Care Policy in Canada, Harvey Lazar, John N. Lavis, Pierre-Gerlier Forest, and John Church (eds.), 2013. ISBN 978-1-55339-324-5

Canada: The State of the Federation 2010, Matthew Mendelsohn, Joshua Hjartarson, and James Pearce (eds.), 2013. ISBN 978-1-55339-200-2

The Democratic Dilemma: Reforming Canada's Supreme Court, Nadia Verrelli (ed.), 2013. ISBN 978-1-55339-203-3